SULFATE REDUCT
REMEDIATION OF GYPSIFEROUS
SOILS AND SOLID WASTES

Thesis Committee

Thesis Promotor

Prof. dr. ir P.N.L. Lens
Professor of Biotechnology
UNESCO-IHE Institute for Water Education
Delft, The Netherlands

Thesis Co-Promotors

Prof. A.P. Annachhatre, PhD
Professor, Environmental Engineering and Management
Asian Institute of Technology
Pathum Thani, Thailand

Dr. Hab. E.D. van Hullebusch, Dr. Hab.,PhD, MSc
Hab. Associate Professor in Biogeochemistry
University of Paris-Est
Paris, France

Dr. G. Esposito, PhD, MSc
Assistant Professor of Sanitary and Environmental Engineering
University of Cassino and Southern Lazio
Cassino, Italy

Other Members

Dr. E. Şahinkaya, PhD
Istanbul Medeniyet University
Istanbul, Turkey

Dr. ir. S. Hiligsmann
University of Liège
Liège, Belgium

This research was conducted under the auspices of the Erasmus Mundus Joint Doctorate Environmental Technologies for Contaminated Solids, Soils, and Sediments (ETeCoS^3) and The Netherlands Research School for the Socio-Economic and Natural Sciences of the Environment (SENSE).

Joint PhD degree in Environmental Technology

UNIVERSITÉ
—PARIS-EST

Docteur de l'Université Paris-Est
Spécialité : Science et Technique de l'Environnement

Dottore di Ricerca in Tecnologie Ambientali

UNESCO-IHE
Institute for Water Education

Degree of Doctor in Environmental Technology

Thèse – Tesi di Dottorato – PhD thesis

Pimluck Kijjanapanich

Sulfate Reduction for Remediation of Gypsiferous Soils and Solid Wastes

To be defended 18th November 2013

In front of the PhD committee

Dr. ir. S. Hiligsmann	Reviewer
Dr. E. Şahinkaya, PhD	Reviewer
Prof. dr. ir P.N.L. Lens	Promotor
Prof. A.P. Annachhatre, PhD	Co-Promotor
Dr. G. Esposito, PhD, MSc	Co-Promotor
Dr. Hab. E.D. van Hullebusch, PhD, MSc	Co-Promotor
Prof. dr. habil. S. Uhlenbrook	Vice Rector of UNESCO-IHE

ERASMUS
MUNDUS

Erasmus Joint doctorate programme in Environmental Technology for Contaminated Solids, Soils and
Sediments (ETeCoS[3])

CRC Press/Balkema is an imprint of the Taylor & Francis Group, an informa business

© 2013, Pimluck Kijjanapanich

Published by:
CRC Press/Balkema
PO Box 11320, 2301 EH Leiden, The Netherlands
e-mail: Pub.NL@taylorandfrancis.com
www.crcpress.com – www.taylorandfrancis.com

ISBN: 978-1-138-01535-7

Table of Contents

Acknowledgements

I would like to express my sincere gratitude to Prof. dr. ir. P.N.L. Lens, my promotor for his invaluable suggestions, help and encouragement throughout my studies and thesis work.

I would like to thank my co-promotors, Prof. A.P. Annachhatre, Dr. G. Esposito and Dr. Hab. E.D. van Hullebusch, for their great support and scientific contribution for my research. My grateful acknowledgement is further extended to my thesis examination committee members, Dr. E. Şahinkaya and Dr. ir. S. Hiligsmann, for their valuable comments and suggestions.

I am highly grateful to my scholarship donor, the Erasmus Mundus Joint Doctorate Environmental Technologies for Contaminated Solids, Soils, and Sediments (ETeCoS^3) (FPA n°2010-0009), who provide such a special opportunity to study at UNESCO-IHE Institute for Water Education (The Netherlands), University of Cassino and Southern Lazio (Italy) and Asian Institute of Technology (Thailand). This research was also supported through the UNESCO-IHE Partnership Research Fund (UPaRF) (The Netherlands) in the Permeable Reactive Barrier for Remediation of Acid Mine Drainage project (60042391 UPaRF PRBRAMD Type III).

Sincere thanks also pass through Mrs. B.P. Pedrajas, Dr. E.R. Rene, the entire faculty, staff, laboratory supervisors, laboratory technicians of UNESCO-IHE who were always helpful. The authors would like to thank the Laboratoire Géomatériaux et Environnement of the Université Paris-Est for the analytical support.

The authors also sincerely thank Smink Afvalverwerking B.V. (Amersfoort, The Netherlands), Membranes International Inc. (USA), Hoogh Electronic Components B.V. (Delft, The Netherlands) and the Energy Research and Development Institute-Nakonping, Chiang Mai University (Chiang Mai, Thailand), which provided the construction and demolition debris, the cation exchange membrane, the resistor and seed sludge for this study In addition, thanks are also due to all my dear friends for their encouragement and help.

Finally, I would like to express my thanks to my respected parents who gave invaluable helpful suggestion, support, cheer and help.

Summary

Solid wastes containing sulfate, such as construction and demolition debris (CDD), are an important source of pollution, which can create a lot of environmental problems. It is suggested that these wastes have to be separated from other wastes, especially organic waste, and place it in a specific area of the landfill. This results in the rapid rise of the disposal costs of these gypsum wastes. Although these wastes can be reused as soil amendment or to make building materials, a concern has been raised by regulators regarding the chemical characteristics of the material and the potential risks to human health and the environment due to CDD containing heavy metals and a high sulfate content.

Soils containing gypsum, namely gypsiferous soils, also have several problems during agricultural development such as low water retention capacity, shallow depth to a hardpan and vertical crusting. In some mining areas, gypsiferous soil problems occur, coupled with acid mine drainage (AMD) problems which cause a significant environmental threat. Reduction of the sulfate content of these wastes and soils is an option to overcome the above mentioned problems. This study aimed to develop sulfate removal systems to reduce the sulfate content of CDD and gypsiferous soils in order to decrease the amount of solid wastes as well as to improve the quality of wastes and soils for recycling purposes or agricultural applications.

The treatment concept leaches the gypsum contained in the CDD by water in a leaching step. The sulfate containing leachate is further treated in biotic or abiotic systems. Biological sulfate reduction systems used in this research were the Upflow Anaerobic Sludge Blanket (UASB) reactor, Inverse Fluidized Bed (IFB) Reactor and Gas Lift Anaerobic Membrane Bioreactor (GL-AnMBR). The highest sulfate removal efficiency achieved from these three systems ranges from 75 to 95%. The treated water from the bioreactor can then be reused in the leaching column. Chemical sulfate removal (abiotic system) is an alternative option to treat the CDD leachate. Several chemicals were tested including barium chloride, lead(II) nitrate, calcium chloride, calcium carbonate, calcium oxide, aluminium oxide and iron oxide coated sand. A sulfate removal efficiency of 99.9% was achieved with barium chloride and lead(II) nitrate.

For AMD and gypsiferous soils treatment, five types of organic substrate including bamboo chips (BC), municipal wastewater treatment sludge (MWTS), rice husk (RH), coconut husk chip (CHC) and pig farm wastewater treatment sludge (PWTS) were tested as electron donors for biological sulfate reduction treating AMD. The highest sulfate reduction efficiency (84%) was achieved when using the combination of PWTS, RH and CHC as electron donors. Then, this organic mixture was further used for treatment of the gypsiferous soils. The gypsum mine soil (overburden) was mixed with an organic mixture in different amounts including 10, 20, 30 and 40% of soil. The highest sulfate removal efficiency of 59% was achieved in the soil mixture which contained 40% organic material.

The removal of sulfide from the effluent of the biological sulfate reduction process is required as sulfide can cause several environmental impacts or be re-oxidized to sulfate if directly discharged to the environment. Electrochemical treatment is one of the alternatives for sulfur recovery from aqueous sulfide. A non-catalyzed graphite electrode was tested as electrode for the electrochemical sulfide oxidation. A high surface area of the graphite electrode is required in order to have less internal resistance as much as possible. The highest sulfide oxidation rate was achieved when using the external resistance at 30 Ω at a sulfide concentration of 250 mg L^{-1}.

Résumé

Les déchets solides contenant des sulfates, comme les déchets de la construction (DC), sont une source importante de pollution susceptible de créer beaucoup de problèmes environnementaux. Il est suggéré que ces déchets doivent être séparés des autres déchets, notamment les déchets organiques, et de les placer dans une zone spécifique de la décharge. Cela se traduit par l'augmentation rapide des coûts d'élimination de ces déchets contenant du gypse. Bien que ces déchets peuvent être réutilisés comme amendement de sol ou de faire des nouveaux matériaux de construction, un problème a été soulevé par le législateur en ce qui concerne les caractéristiques chimiques des déchets de la construction et les risques potentiels pour la santé humaine et l'environnement, en raison de leurs teneurs en métaux lourds et d'une teneur élevée en sulfates.

Les sols contenant du gypse, à savoir les sols gypsifères, engendrent également des problèmes au cours de leur exploitation agricole tels que la faible capacité de rétention d'eau et la formation de croûtes cuirassées. Dans certaines zones minières, les problèmes du sol gypsifères sont associés à la présence de drainages miniers acides (DMA) qui engendre une menace environnementale importante. La réduction de la teneur en sulfates de ces déchets et sols est une option pour surmonter les problèmes mentionnés ci-dessus. Ce travail de thèse visait à développer des procédés d'élimination des sulfates permettant la réduction des teneurs en sulfates des DC et des sols gypsifères afin d'améliorer la qualité des déchets et des sols à des fins agricoles ou des applications de recyclage.

Le concept de traitement des DC par lixiviation à l'eau a été étudié (colonne de lixiviation). Les sulfates contenus dans les lixiviats sont ensuite éliminés à l'aide d'un traitement chimique ou biologique. L'approche biologique mise en oeuvre dans ce travail a consisté à mettre en oeuvre la réduction biologique des sulfates au sein de bioréacteurs de conception différente (i.e. réacteur UASB, réacteur à lit fluidisé inverse (IFB) ou d'un réacteur anaérobie gas lift). L'efficacité d'élimination des sulfates la plus élevée atteinte par ces trois systèmes varie de 75 à 95%. L'eau traitée provenant du bioréacteur peut alors ensuite être réutilisé dans la colonne de lixiviation. Le traitement chimique des sulfates est une option alternative pour traiter les lixiviats. Plusieurs produits chimiques ont été testés, (chlorure de baryum, nitrate de plomb (II), le chlorure de calcium, le carbonate de calcium, l'oxyde de calcium, et du sable recouvert d'un mélange d'oxydes d'aluminium et de fer). Un rendement de 99,9% d'élimination des sulfates (par précipitation) a été atteint avec le chlorure de baryum et le nitrate de plomb (II).

Pour le traitement des DMA et des sols gypseux, cinq types de substrat organique tel que les copeaux de bambou, les boues d'épuration des eaux usées municipales, de l'écorce de riz, de coques de noix de coco broyée et des boues d'épuration des eaux usées d'une ferme porcine ont été testés comme donneurs d'électrons pour la réduction biologique des sulfates. L'efficacité de la réduction des sulfates la plus élevé (84%) a été obtenue en utilisant un mélange d'écorce de riz, de coques de noix de coco broyée et des boues d'épuration des eaux usées d'une ferme porcine comme donneurs d'électrons. Ensuite, ce mélange organique a été utilisé pour le traitement des sols gypsifères. Le sol de la mine de gypse a été mélangé avec le mélange organique en différentes proportions

(10, 20, 30 et 40% de sol). Le rendement le plus élevé de 59% de réduction des sulfates a été atteint dans le mélange de sol qui contient 40% de matière organique.

L'élimination des sulfures présents dans l'effluent des procédés de réduction biologique des sulfates est nécessaire. En effet, les sulfures peuvent causer plusieurs impacts environnementaux ou être ré-oxydé en sulfate si ils sont directement rejetés dans l'environnement. Le traitement électrochimique des effluents est l'une des solutions alternatives pour la récupération du soufre élémentaire à partir des sulfures. Une électrode de graphite a été testée comme électrode permettant l'oxydation électrochimique des sulfures en soufre élémentaire. Une électrode en graphite de grande surface est nécessaire afin d'avoir une résistance électrique la plus faible possible. La vitesse d'oxydation des sulfures la plus élevée est atteinte lors de l'application d'une résistance de 30 Ω à une concentration en sulfure de 250 mg.L^{-1}.

Sommario

I rifiuti solidi contenenti solfati, come i detriti da costruzione e demolizione (CDD), sono un importante fonte d'inquinamento, che può creare molti problemi ambientali. È consigliabile separare questi rifiuti dagli altri, specialmente i rifiuti organici, e posizionarli in aree specifiche delle discariche. Questo determina una rapida crescita dei costi di smaltimento di questi rifiuti contenti gesso. Nonostante questi rifiuti possano essere riutilizzati come ammendanti per il suolo o per realizzare materiali da costruzione, le normative ambientali hanno sollevato il problema relativo alle caratteristiche chimiche del materiale e ai potenziali rischi per la salute umana e l'ambiente legati al fatto che i CDD contengono metalli pesanti e presentano un alto contenuto di solfati.

I suoli contenenti gesso, detti suoli gessiferi, presentano anche diversi problemi per lo sviluppo agricolo, come la bassa capacità di ritenzione idrica, la bassa profondità e l'incrostazione verticale. In alcune aree minerarie, inoltre, si verificano problemi legato alla presenza combinata di suoli contenti gesso e scarico di acque acide, che causano serie minacce ambientali. La riduzione del contenuto di solfato di questi rifiuti e suoli è un opzione per superare i suddetti problemi. Questo studio è stato mirato a sviluppare sistemi di rimozione dei solfati da CDD e suoli gessiferi, per ridurre la quantità di rifiuti solidi prodotti e migliorare la qualità dei rifiuti e dei suoli per fini di riciclo e applicazioni agricole.

Il trattamento proposto consiste nella lisciviazione con acqua del gesso contenuto nei CDD. Il solfato contenuto nel percolato viene ulteriormente trattato in sistemi biotici e abiotici. I sistemi biologici di riduzione del solfato utilizzati in questa ricerca sono stati i seguenti: *Up-flow Anaerobic Sludge Blanket (UASB) Reactor*, *Inverse Fluidized Bed (IFB) Reactor* e *Gas Lift Anaerobic Membrane Bioreactor (GL-AnMBR)*. Le massime efficienze di rimozione del solfato ottenute con questi tre sistemi variano tra 75 e 95%. L'acqua trattata dal bioreattore può poi essere riutilizzata nella colonna di lisciviazione. La rimozione chimica del solfato (sistema abiotico) è un'opzione alternativa per il trattamento del percolato dei CDD. Diverse sostanze chimiche sono state testate compreso cloruro di bario, nitrato di piombo (II), cloruro di calcio, carbonato di calcio, ossido di calcio, ossido di alluminio e sabbia rivestita con ossido di ferro. Con cloruro di bario e nitrato di piombo (II) è stata raggiunta un'efficienza di rimozione del solfato del 99,9%.

Per il trattamento combinato di AMD e suoli gessiferi, cinque tipi di substrato organico sono stati testati come donatori di elettroni per la riduzione dei solfati nel trattamento dell'AMD, vale a dire cippati di bambù (BC), fanghi di trattamento di acque reflue municipali (MWT), lolla di riso (RH), cippato di buccia di cocco (CHC) e fanghi della depurazione di reflui suinicoli (PWTS). La massima efficienza di riduzione del solfato (84%) è stata ottenuta usando la combinazione di PWTS, RH e CHC come donatori di elettroni. Pertanto questa miscela organica è stata ulteriormente utilizzata per il trattamento dei suoli gessiferi. Campioni di suolo prelevati in cave di gesso sono stati miscelati con la miscela organica in percentuali differenti, tra cui 10, 20, 30 e 40% di suolo. La massima efficienza di rimozione del solfato, pari al 59%, è stata raggiunta con una miscela che conteneva il 40% di materiale organico.

La rimozione del solfuro dall'effluente del processo biologico di riduzione dei solfati è necessaria poiché il solfuro può causare diversi impatti ambientali o essere ri-ossidato a solfato se scaricato direttamente nell'ambiente. Il trattamento elettrochimico è una delle alternative per il recupero dello zolfo dal solfuro in fase acquosa. Un elettrodo di grafite non catalizzato è stato testato come elettrodo per l'ossidazione elettrochimica del solfuro. E' richiesta un'elevata area superficiale dell'elettrodo di grafite per avere una minore resistenza interna. E' stato raggiunto il più alto tasso di ossidazione di solfuro quando si è utilizzata la resistenza esterna a 30 Ω ad una concentrazione di solfuro di 250 mg L^{-1}.

Samenvatting

Vast afval vervuild met sulfaat, zoals bouw- en sloopafval (BSA), zijn belangrijke bronnen van vervuiling die een aantal milieuproblemen kunnen veroorzaken. Dit afval wordt bij voorkeur gescheiden van andere afvalsoorten, in het bijzonder organisch afval, en op een aparte plaats op een stortplaats opgeslagen. Dit leidt tot een fikse toename in de stortkosten van gipshoudend afval. Hoewel dit afval kan worden hergebruikt als bodemverbeteraar of als bouwmateriaal, zijn er door de wetgeving beperkingen met betrekking tot de chemische eigenschappen van het materiaal en de potentiële risico's voor de volksgezondheid en het milieu door de verontreiniging van BSA door zware metalen en hoge sulfaat gehaltes.

Gipshoudende bodems hebben ook diverse problemen bij de landbouwkundige ontwikkeling zoals een lage water retentie, lage diepte tot de hardpan en verticale korstvorming. In sommige mijngebieden komen problemen met gipshoudende bodems voor, gekoppeld met de vorming van zuur mijndrainage water kan dit een significante bedreiging voor het milieu betekenen. Een reductie van het sulfaat gehalte van deze afval- en bodemtypes is een mogelijkheid om bovenvermelde problemen op te lossen. Deze studie beoogde om een sulfaat verwijderingsystemen te ontwikkelen om het sulfaat gehalte van BSA en gipshoudende bodems alsook de hoeveelheid BSA afval te verminderen en de kwaliteit van gipshoudende bodems te verbeteren voor recyclage doeleinden of landbouwkundige toepassingen.

Het behandelingsconcept loogt het gips bevat in het BSA uit met water in een uitloogstap. Het sulfaathoudend leachate wordt verder behandeld in een biotisch of abiotisch systeem. Biologische sulfaat reductie systemen gebruikt in dit onderzoek zijn de Opstoom Anaerobe Slib Bed (UASB) reactor, de Inverte Fluidized Bed (IFB) Reactor en de Gas Lift Anaerobe Membraan Bioreactor (GL-AnMBR). De hoogst bereikte sulfaat verwijderingefficiëntie voor deze drie systemen bedroeg 75 to 95%. Het behandelde water van deze bioreactoren kan hergebruikt worden in de uitloogkolom. Chemische sulfaat verwijdering (abiotisch system) is een alternatieve optie om BSA uitloogwater te behandelen. Verscheidene chemiciën werden getest, inclusief barium chloride, lood(II)nitraat, calcium chloride, calcium carbonaat, calcium oxide, aluminium oxide and ijzer oxide gecoat zand. Een sulfaat verwijderingefficiëntie van 99.9% werd bereikt met barium chloride en lood (II) nitraat.

Voor de behandeling van zuur mijndrainage water en gipshoudende bodems werden vijf types organische substraten, met name bamboe chips, huishoudelijk afvalwater behandelingsslib (MWTS), rijst kaf (RK), kokosnoot kaf chips en varkenshouderij afvalwater behandelingsslib (VABS), getest als elektron donor voor de biologische sulfaatreductie. Het hoogste sulfaat reductie rendement (84%) werd behaald met de combinatie van VABS, RK en CKC als elektron donor. Dan werd dit organisch mengsel getest voor de behandeling van gipshoudende bodems. De gipshoudende bodem werd gemengd met het organisch mengsel in verschillende hoeveelheden inclusief 10, 20, 30 en 40% van de bodem. De hoogste sulfaat verwijderingefficiëntie van 59% werd bekomen in het bodemmengsel die 40% organisch materiaal bevatte.

De verwijdering van sulfide van het effluent van het biologische sulfaat reductie proces is nodig omdat sulfide verschillende milieu-impacts heeft of terug geoxideerd kan worden tot sulfaat. Elektrochemische behandeling is één van de alternatieven die ook zwavel hergebruik van opgelost sulfide mogelijk maakt. Een grafiet elektrode werd getest als elektrode voor elektrochemische sulfide oxidatie. Een hoog oppervlak van de grafietelektrode is nodig om zo min mogelijk interne weerstand te hebben. De hoogste sulfide oxidatie snelheid werd behaald bij een externe weerstand van 30 Ω bij een sulfide concentratie van 250 mg L^{-1}.

CHAPTER 1

Introduction

Chapter 1

Gypsum or calcium sulfate dihydrate ($CaSO_4 \cdot 2H_2O$) is nontoxic to humans and can be helpful to animals and plant life. Gypsum is mined, processed and converted into several products. It can be used in agriculture as an amendment, conditioner, as well as fertilizer. This is because gypsum can improve water penetration, be used for reclamation of sodic soils, help plants absorb plant nutrients, stop water runoff and erosion. However, if soils contain too much gypsum (more than 10%), growth of plant roots becomes inhibited and gypsum precipitation increasingly tend to break the continuity of the soil mass (Verheye & Boyadgiev, 1997). Soils that contain sufficient quantities of gypsum to interfere with plant growth and crop production are called "gypsiferous soils" (FAO, 1990).

Gypsum is also widely used in the construction industry and it is a major component in drywalls (gypsum boards). However, construction, renovation or demolition activities yield large amounts of gypsum contaminated wastes called construction and demolition debris (CDD). It is also produced in large quantities as a by-product from fertilizer manufacturing or as desulfurization product during the treatment of waste gases from coal combustion processes. These solid by-products become gypsum waste, resulting into large quantities of waste due to industrial growth.

1.1. Problem Description

The presence of gypsum in gypsiferous soils creates several problems for their agricultural use and development, including low water retention capacity, shallow depth to the hardpan and vertical crusting (Khresat et al., 2004). The accumulation of gypsum in soils results in very low fertility, and consequently, their productivity remains low under irrigation even with application of fertilizers or organic manures (FAO, 1990). These problems also occur in several mining areas, especially gypsum mines, where the soils have a high gypsum content and cannot be used for agriculture. For instance, soils in the gypsum mine area in the southern part of Thailand have a high sulfate content that can induce adverse effects on the environment. Moreover, the soils of some mines can also generate acid mine drainage (AMD) and mass mortalities of plants and aquatic life (Kijjanapanich et al., 2012). This AMD has a low pH and high concentrations of sulfate and toxic metals. Such land cannot be used for agriculture, and these soils have a poor fauna and flora.

Construction, renovation or demolition activities yield large amounts of CDD. Nearly 40% of the total mass of CDD consists of a fine fraction containing high amounts of gypsum (Montero et al., 2010; Townsend et al., 2004), namely CDD sands (CDDS). For applications where the CDD is placed in direct contact with the environment, there are potential regulatory concerns regarding the high levels of sulfate and heavy metals in CDD and the potential risks to human health and the environment (Jang & Townsend, 2001).

The Dutch government has set the limits to the maximum amount of polluting compounds present in building material. For reusable sand, the emission limit is 1.73 g sulfate per kg of sand (de Vries, 2006; Stevens, 2013). Therefore, most of the CDD cannot be reused for construction activities due to its high sulfate content. Moreover, deposition of CDD in landfills can lead to exceptionally high levels of biogenic sulfide

(H$_2$S), posing serious problems such as odor (Jang, 2000; Lens & Kuenen, 2001), pipe corrosion (Vincke et al., 2001) and contamination of landfill gas (Karnachuk et al., 2002) or groundwater. Thus, gypsum waste has to be separated from other wastes, especially organic waste, and placed in a specific area of a landfill. This results in a rise of the disposal costs of gypsum waste (Gypsum Association, 1992).

Reduction of the sulfate content of these wastes and soils is an option to overcome the above mentioned problems. Remediation of gypsiferous soils in the abandoned mine areas seem to be a win-win solution for both waste lands (uncultivated lands) and AMD problems. If these problems are solved, agricultural areas of the world could be increased. In addition, these abandoned lands can be used not only for cultivation, but also for reforestation that can reduce greenhouse effects or global warming. In case of the CDD, not only the amount of solid wastes can be reduced, but also the treated CDD and sulfur can be reused and recovered.

This study aimed to develop sulfate removal systems to reduce the sulfate content of CDD and gypsiferous soils in order to decrease the amount of solid wastes as well as to improve the quality of wastes and soils for recycling purposes or agricultural applications.

1.2. Objectives

The main objective of this research is *"to develop an appropriate system for sulfate removal from gypsiferous soils and solid wastes"*.

The specific objectives are:

1). To study the characteristics of gypsiferous soils and solid wastes (CDD).
   a. To investigate physical and chemical characteristics of gypsiferous soils
   b. To investigate the leaching potential of gypsum from gypsiferous soils and CDD
2). To study the *in situ* biological sulfate reduction for sulfate removal from AMD and gypsiferous soils.
   a. To select an appropriate organic material used as electron donor for treating AMD using permeable reactive barriers (PRB)
   b. To investigate the appropriate ratio of organic material (electron donor) to gypsiferous soils for gypsiferous soils remediation using SRB
   c. To investigate the optimum residence time for achieving AMD and gypsiferous soils treatment using biological sulfate reduction processes
3). To study the *ex situ* sulfate reduction system for sulfate removal from CDD.
   a. To develop a biological sulfate removal system to reduce the sulfate content of CDD and to treat CDD leachate for reuse in the leaching column
   b. To investigate the effect of the calcium concentration contained in the CDD leachate on the biological sulfate removal efficiency
   c. To study chemical sulfate removal as an alternative for sulfate removal from sulfate CDD leachate
4). To study the electrochemical treatment as an alternative option for elimination of sulfide generated from biological sulfate reduction process.
   a. To investigate the appropriate external resistance for elimination of sulfide generated from the biological sulfate reduction process

b. To investigate the effect of internal resistance on the sulfide removal efficiency and electrical current

c. To study the sulfide removal efficiency using electrochemical treatment at different pH values

1.3. Structure of Thesis

The present dissertation comprises nine chapters. The following paragraphs outline the content of the chapters (Figure 1.1).

Chapter 1 gives a general overview of the research, including background, problem description, research objectives and thesis structure.

Chapter 2 gives a literature review about the problem related to sulfate rich soils, sediments and solid wastes, presents their characteristics and overview current methods of their bioremediation.

Chapter 3 presents an investigation of using low or no cost organic substrates as electron donor for SRB in a biological sulfate reducing PRB, in order to remove sulfate and heavy metals from AMD.

Chapter 4 investigates the characteristics of mine soils from Thailand and the treatment of gypsiferous mine soils by biological sulfate reduction using organic substrates as electron donors for SRB.

Chapter 5 develops a biological sulfate removal system to reduce the sulfate content of CDD. The leachability of CDD gypsum in a leaching column was also investigated.

Chapter 6 compares the treatment of CDD leachate using three different types of bioreactor, including the Upflow Anaerobic Sludge Blanket (UASB) reactor, Inverse Fluidized Bed (IFB) reactor and Gas Lift Anaerobic Membrane Bioreactor (GL-AnMBR). The effect of the calcium concentration contained in the CDD leachate on the sulfate removal efficiency was also investigated.

Chapter 7 presents the use of the chemical sulfate removal as an alternative for sulfate removal from CDD leachate. Both sulfate precipitation and adsorption were investigated to find an appropriate chemical sulfate removal process.

Chapter 8 explores the electrochemical treatment for treating the effluent from a sulfate reducing bioreactor by using the spontaneous electrochemical sulfide oxidation/vanadium(V) reduction with graphite electrode. The effect of both internal and external resistance, and pH on the sulfide removal efficiency and electrical current were also investigated.

Chapter 9 summarizes and draws conclusions on knowledge gained from this study and gives recommendations for future perspective.

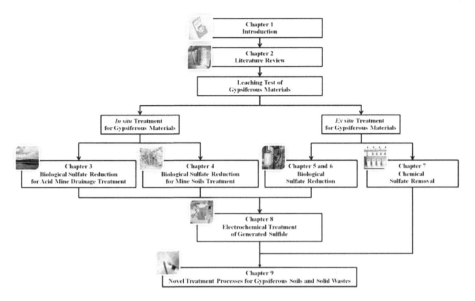

Figure 1.1. Overview of thesis

1.4. References

de Vries, E. (2006). *Biological sulfate removal from construction and demolition debris sand.* Wageningen University.

FAO. (1990). *FAO Soils Bulletin 62: Management of gypsiferous soils.* Rome.

Gypsum Association. (1992). Treatment and disposal of gypsum board waste: Technical paper part II, *AWIC's Construction Dimensions* (Vol. March): AWIC.

Jang, Y. (2000). *A study of construction and demolition waste leachate from laboratory landfill simulators.* University of Florida, Florida.

Jang, Y. C., & Townsend, T. (2001). Sulfate leaching from recovered construction and demolition debris fines. *Adv. Environ. Res., 5,* 203-217.

Karnachuk, O. V., Kurochkina, S. Y., & Tuovinen, O. H. (2002). Growth of sulfate-reducing bacteria with solid-phase electron acceptors. *Appl. Microbiol. Biotechnol., 58,* 482-486.

Khresat, S. A., Rawajfih, Z., Buck, B., & Monger, H. C. (2004). Geomorphic features and soil formation of arid lands in Northeastern Jordan. *Arch. Agron. Soil Sci., 50,* 607-615.

Kijjanapanich, P., Pakdeerattanamint, K., Lens, P. N. L., & Annachhatre, A. P. (2012). Organic substrates as electron donors in permeable reactive barriers for removal of heavy metals from acid mine drainage. *Environ. Technol., 33*(23), 2635-2644.

Lens, P. N. L., & Kuenen, J. G. (2001). The biological sulfur cycle: novel opportunities for environmental biotechnology. *Water Sci. Technol., 44*(8), 57-66.

Montero, A., Tojo, Y., Matsuto, T., Yamada, M., Asakura, H., & Ono, Y. (2010). Gypsum and organic matter distribution in a mixed construction and demolition waste sorting process and their possible removal from outputs. *J. Hazard. Mater., 175,* 747-753.

Stevens, W. (2013). Personal Communication www.smink-groep.nl.

Townsend, T., Tolaymat, T., Leo, K., & Jambeck, J. (2004). Heavy metals in recovered fines from construction and demolition debris recycling facilities in Florida. *Sci. Total Environ., 332*, 1-11.

Verheye, W. H., & Boyadgiev, T. G. (1997). Evaluating the land use potential of gypsiferous soils from field pedogenic characteristics. *Soil Use Manage., 13*, 97-103.

Vincke, E., Boon, N., & Verstraete, W. (2001). Analysis of the microbial communities on corroded concrete sewer pipes - a case study. *Appl. Microbiol. Biotechnol., 57*, 776-785.

CHAPTER 2

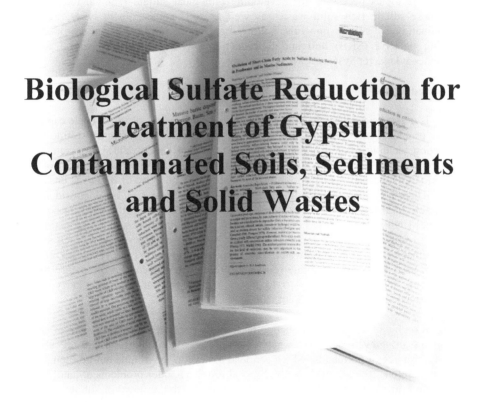

Biological Sulfate Reduction for Treatment of Gypsum Contaminated Soils, Sediments and Solid Wastes

This chapter has been published as:
Kijjanapanich, P., Annachhatre, A. P., & Lens, P. N. L. (2013). Biological sulfate reduction for treatment of gypsum contaminated soils, sediments and solid wastes. *Crit. Rev. Environ. Sci. Technol., In Press*. DOI: 10.1080/10643389.2012.743270

Chapter 2

Solid wastes containing sulfate are an important source of pollution, which can create a lot of environmental problems, especially during disposal management at landfill sites. These solid wastes, such as construction and demolition debris (CDD) and phosphogypsum, cause odor problems and possible health impacts to landfill employees and surrounding residents. These wastes do not only contain high sulfate concentrations, but also contain toxic metals and radioactive compounds. Although these wastes can be reused as soil amendment or to make building materials, a concern has been raised by regulators regarding to the chemical characteristics of the material and the potential risks to human health and the environment. Therefore, use of these solid wastes has been banned in most countries. In addition, soils containing solid sulfate (gypsum), namely gypsiferous soils, have several problems during agricultural development. Reduction of the sulfate content of these solid wastes, soils and sediments by biological sulfate reduction is an option to overcome the above mentioned problems. This paper reviews the topics necessary for developing biological sulfate removal technologies from these sulfate rich solid wastes as well as soils and sediment types, i.e. their contamination by sulfate minerals, solid sulfate as an electron acceptor for sulfate reducing bacteria (SRB) and sulfate reduction processes both in natural and in bioengineered reactor systems.

2.1. Introduction

Sulfate is a nontoxic ion but its conversions within the biological sulfur cycle can cause several problems affecting the environment. These include hydrogen sulfide (H_2S) production yielding toxicity, odor problems (Lens & Kuenen, 2001), concrete sewer pipe corrosion (Vincke et al., 2001), increase of the liquid effluent chemical oxygen demand (COD) as well as deterioration in quality and quantity of biogas (Lens et al., 1998). Therefore, biological sulfate reduction in the past has been considered as undesirable in anaerobic wastewater treatment (Hulshoff Pol et al., 1998). In contrast, from the 1990's, interest has grown in applying biological sulfate reduction for treatment of specific waste streams (inorganic sulfate rich wastewaters), such as acid mine drainage (AMD) or wastewater containing sulfuric acid (Lens et al., 2002). This approach uses the bacterial sulfate reduction process as it occurs in the nature for the removal of sulfate, often coupled to heavy metal removal (Jong & Parry, 2003; Kijjanapanich et al., 2012; Liamleam, 2007).

Research on biological sulfate reduction has mainly focused on the treatment of sulfate containing groundwater or wastewaters. Solid wastes containing sulfate are also an important source of pollutants, which can lead to several environmental problems upon its reduction to sulfide, such as waste disposal problems, odor problems at landfill sites and groundwater contamination. A novel approach for the removal of sulfate has been developed for the treatment of sulfate containing wastewaters which can also be applied to soils, sediments and solid wastes. There is an increasing interest in biotechnological applications using SRB as an alternative method for sulfate and heavy metal removal from environmental contamination (Chang et al., 2000; Elliott et al., 1998). Sulfate removal is not only capable of solving these problems, but sulfide produced in this process can also be recovered back as elemental sulfur (S^0) and the amount of solid waste which needs to be disposed to landfill sites is reduced. This review addresses problems related to sulfate rich soils, sediments and solid wastes, presents their characteristics and overviews methods of their bioremediation.

2.2. Soils, Sediments and Solid Wastes Contaminated by Solid Sulfur
2.2.1 Soils

Gypsiferous soils are those which contain significant quantities of gypsum ($CaSO_4 \cdot 2H_2O$) which may interfere in plant growth (FAO, 1990). Gypsum can be transported by water or wind and re-deposited at new locations forming individual gypsum dunes or becoming incorporated in the soil. The main reason for gypsum accumulation in the soil is its precipitation from underground and runoff waters, as a result of intensive evaporation. In addition, the origin of the sulfate ions (SO_4^{2-}) in the soil solution is due to the presence of sulfur rich minerals such as pyrite (FeS_2) in the parent material. By weathering and oxidation, the sulfur in these minerals is transformed into sulfuric acid, which in calcareous soils reacts with calcium carbonate ($CaCO_3$) to form gypsum (FAO, 1990).

Gypsiferous soils cover about 94 million ha of the world's arable lands (FAO, 1993). These soils are predominantly present in dry areas (with less than 400 mm annual rainfall), where sources for calcium sulfate exist (Porta & Herrero, 1990). The agricultural utilization of gypsiferous soils is limited due to the presence of gypsum that can induce hardpan formation and vertical crusting. The accumulation of gypsum in soils results in very low fertility, and consequently, their productivity remains low under irrigation even with application of fertilizers or organic manures (FAO, 1990). The physical structure such as porosity and permeability of gypsiferous soils can be improved by reducing the soil's gypsum content (Alfaya et al., 2009).

Some abandoned mine areas, especially gypsum mines, are also a source of gypsiferous soils. For instance, the overburden in the abandoned gypsum mine in Surat Thani, Thailand (Figure 2.1a) has a high sulfate content that can contaminate the environment. Moreover, the soil in this area also contains pyrite, resulting in formation of AMD. AMD has a low pH and high concentrations of sulfate as well as toxic metals. This land cannot be used for agriculture, and there are very few plants and animals.

The presence of gypsum in gypsiferous soils creates several problems for their agricultural development, including low water retention capacity, shallow depth to a hardpan and vertical crusting (Khresat et al., 2004). Compared to a non-gypsiferous soil, the activities of the calcium and sulfate ions in the soil solution are increased due to the solubility of gypsum, resulting in the common ion effect which may cause calcite to precipitate (Kordlaghari & Rowell, 2006). A gypsum content of 3-10% does not interfere significantly with soil characteristics such as its structure. The gypsum crystals however tend to break the continuity of the soil mass in soils containing 10-25% of gypsum. The soils with more than 25% of gypsum do not provide a good medium for plant growth. Under such conditions, gypsum may precipitate and can cement soil material into hard layers, thus reducing root penetration and causing plant cultivation problems (Smith & Robertson, 1962).

Figure 2.1. Gypsum containing materials:
(a) gypsiferous soil in gypsum mine, (b) phosphogypsum stacks, (c) construction and demolition debris (CDD) and (d) flue gas desulfurization (FGD) gypsum.

2.2.2 Sediments

The majority of the sulfur in lakes and rivers originates from the weathering of sulfur containing rocks in the catchment and from the oxidation of organic sulfur from terrestrial sources. However, at present a large proportion of sulfur comes from the burning of fossil fuels and discharge of sulfate containing wastewater (Dornblaser et al., 1994) or solid waste into water (Lloyd, 1985). Higher sulfate/sulfite atmospheric concentrations in acid rain and the discharge of wastewater with a high sulfate concentration affect the sulfur cycling in lakes and rivers (Peiffer, 1998). Many lakes have changed from oligotrophic to meso- or eutrophic conditions during the past decades because of nutrient loading from wastewater and fertilizers (Holmer & Storkholm, 2001). This has caused a significant increase in the availability of electron acceptors, such as sulfate and nitrate, in many freshwater wetlands and resulted in severe problems for the freshwater wetlands (Lamers et al., 2001; Lamers et al., 1998).

The annual deposition of organic material on the sea floor is about ten billion tons (Jørgensen & Kasten, 2006). As the particulate organic matter is deposited on the sea floor, it is immediately attacked by a broad range of organisms that all contribute to its degradation and gradual mineralization (Jørgensen & Kasten, 2006). Seawater contains around 28 mM of sulfate. Therefore, organic matter oxidation in marine sediments is an important part coupled to sulfate reduction (Meulepas et al., 2010).

The sulfur cycling in aquatic sediments involves both reductive and oxidative processes (Jørgensen, 1990) and it is both spatially and temporally dynamic. It also strongly influences many biogeochemical reactions in sediments, such as the binding of phosphorus (Lamers et al., 1998). An increase in sulfate availability in freshwater will stimulate sulfate reduction in soils and sediments. Sulfide produced by sulfate reduction interferes with the iron-phosphate precipitates in soils and sediments due to the formation of iron sulfides and associated release of phosphorous (Figure 2.2). Exhaustion of iron from iron sulfide precipitates results in increased sulfide levels and iron shortage in aquatic species, while increased phosphate mobilization and a disturbance of the iron cycle results in increased phosphate levels in the water layer (Smolders & Roelofs, 1993). In this way, the released phosphate causes indirect eutrophication resulting, among others, in a dominance of non-rooting species and algae and, thus increased turbidity of the water.

The Fe:P ratios in the bottom waters of lakes have been found to be significantly related to the surface water sulfate concentrations (Caraco et al., 1993). The higher Fe:P ratios in low sulfate systems is not only due to higher iron concentrations in anoxic bottom waters, but also due to lower P concentrations in anoxic waters (Caraco et al., 1993). Smolders and Roelofs (1993) found that the amount of sulfide accumulating in the sediment highly depends on the availability of soluble iron. Thus, exhaustion of dissolved iron in the sediment parallels both sulfide accumulation and phosphate mobilization.

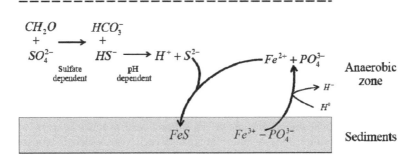

Figure 2.2. Release of phosphate in sediments due to sulfate reduction activity (adapted from Caraco et al., 1993).

2.2.3 Solid wastes

Gypsum is mined and converted into several products, especially useful in construction. It is a major component in drywalls (gypsum board). Gypsum is also produced in large quantities as a by-product from fertilizer manufacturing or as desulfurization product from the coal combustion process. These solid by-products become gypsum waste resulting into large quantities of waste due to industrial growth. In addition, deposition of gypsum containing waste and debris in landfills can lead to exceptionally high levels of biogenic sulfide formation, posing serious problems of odor control and landfill gas purification (Karnachuk et al., 2002). The growing public concern about waste disposal and the environment results a rapid rise in the cost of disposal of gypsum waste (Gypsum Association, 1992b) because these wastes need to be landfilled separately from organic containing wastes. However, some gypsum wastes also contain organic substrates. Therefore, removing organic matter from gypsum rich wastes is necessary before landfilling (Montero et al., 2010).

2.2.3.1 Phosphogypsum

Phosphogypsum is a primary by-product of the phosphate fertilizer industry and emanates from the production of phosphoric acid from phosphate rock (apatite). It is produced from the generation of phosphoric acid by reacting phosphate rock with sulfuric acid according to following equation:

$$Ca_5(PO_4)_3X + 5\ H_2SO_4 + 2\ H_2O \rightarrow 3\ H_3PO_4 + 5\ CaSO_4 \cdot 2\ H_2O + HX \qquad (2.1)$$

where X may include OH⁻, F⁻, Cl⁻, or Br⁻. The calcium sulfate, referred in this context as phosphogypsum, must then be disposed of. The composition of phosphogypsum varies depending on the source of phosphate rock and the phosphoric acid manufacturing process (Mays & Mortvedt, 1986). Table 2.1 shows the composition of some types of phosphogypsum.

In general, phosphogypsum (Figure 2.1b) is a moist, gray, powdery and acidic (pH = 2-5) material containing residual acid, fluoride, toxic metals and radioactive compounds such as uranium and radium those may be present in the phosphate ore. Although the exact quantity produced depends on the phosphate rock source material, the wet-process route produces around five tons of the by-product calcium sulfate per ton phosphorus pentoxide (P_2O_5), the anhydride of phosphoric acid, (Azabou et al., 2005). It is estimated that more than 22 million tons of P_2O_5 are produced annually worldwide (Wissa, 2003), generating around 100-280 million tons of gypsum by-product per year (Tayibi et al., 2009). Since the mid-eighties, the annual production rate of phosphogypsum has been in the range of 40-47 million metric tons per year. The total amount generated in the United States from 1910 to 1981 was about 7.7 billion metric tons. In Central Florida, one of the major phosphoric acid producing areas, industry generates about 32 million tons of phosphogypsum each year which is stockpiled in stacks of nearly 1 billion metric tons (U.S.EPA., 2010).

Table 2.1. Composition of phosphogypsum (The values are relative to phosphogypsum dry weight, % *w/w*)

Component	Tunisia Azabou et al. (2005)	Silesia Wolica and Kowalski (2006)	Texas Taha and Seals (1992)	Florida Taha and Seals (1992)
CaO	30.6	29.6	32.5	25-31
SO$_4$	44.3	50.64	53.1	55-58
P$_2$O$_5$	1-1.5	2.2	0.65	0.5-4.0
Total Kjeldahl Nitrogen	0.076	n.a.	n.a.	n.a.
Fe$_2$O$_3$	0.05	0.14	0.1	0.2
Al$_2$O$_3$	0.11	0.2	0.1	0.1-0.3
SiO$_2$	1.7	0.65	n.a.	n.a.
MgO	0.02	0.05	n.a.	n.a.
Na$_2$O	0.7	0.4	n.a.	n.a.
K$_2$O	0.02	0.1	n.a.	n.a.
Organic carbon	0.45	n.a.	n.a.	n.a.
F	1.3	0.5	1.2	0.2-0.8

n.a.: not available

Phosphogypsum management is one of the most serious problems currently faced by the phosphate industry. Only 15% of the worldwide production is recycled, the remaining 85% is stored without any treatment (Tayibi et al., 2009). This stored phosphogypsum can cause serious environmental problems including soil, water and atmosphere contamination, due to toxic metals and especially radioactive compounds.

There are essentially three methods for disposing of this by-product: discharge into water, dumping on land or utilization as a raw material for chemical manufacturing, agricultural purposes or in construction materials (Lloyd, 1985). Disposing of phosphogypsum by dumping it in water on land-based stacks is widely used, but care must be taken to prevent groundwater contamination. In addition, disposing of gypsum in landfills may lead to exceptionally high levels of biogenic sulfide formation, resulting in, among others, odor problems.

2.2.3.2 Construction and demolition debris (CDD)

CDD (Figure 2.1c) originates from building, demolition and renovation of buildings and roads. With insufficient source separation, CDD becomes a mixed material which is difficult to recycle (Montero et al., 2010). CDD usually contain small pieces of wood, concrete, rock, paper, plastic, metal, and gypsum drywall (Table 2.2). According to several characterization studies of CDD in the US, gypsum drywall accounts for 21-27% of the mass of debris generated during the construction and renovation of residential structures (U.S.EPA., 1998). On average, 0.9 metric tons of waste gypsum is generated from the construction of a typical single family home or 4.9 kg m^{-2} of the structure (Turley, 1998). Nearly 40% of the total mass consists of the fine fraction, called CDD sand (CDDS), which contains high amounts of gypsum (Montero et al., 2010). The content of gypsum (by mass) in CDDS ranges from 1.5% to 9.1% (Jang & Townsend, 2001).

Reuse options have been proposed for CDDS, including soil amendment, alternative daily landfill cover, and fill material in road, embankment and construction projects. The presence of gypsum drywall in CDDS may provide some benefit as a soil conditioner or nutrient source for agriculture. However, for applications where the material is placed in direct contact with the environment, concerns has been raised by regulators regarding the chemical characteristics of the material and the potential risk to human health and the environment (Jang & Townsend, 2001).

Gypsum drywall has been associated with odor problems at many CDD landfills (Jang, 2000). Under extremely wet conditions (high water table), gypsum waste can contribute to the growth of anaerobic bacteria (Gypsum Association, 1992a). When wet landfill conditions occur, it is suggested that this waste be separated from other wastes, especially organic waste, and placed in a specific area of the landfill. This results in the rapid rise of the disposal costs of gypsum waste (Gypsum Association, 1992b).

Montero et al. (2010) found that organic matter was distributed mainly in fractions composed of large-sized components, whereas the gypsum was concentrated in the fine fraction (52.4%). Therefore, the amount of gypsum going to a landfill can be reduced by separating the fine fraction from mixed CDD. However, final disposal still requires removing gypsum also from the fine fraction (CDDS).

Table 2.2. Typical components of construction and demolition debris (CDD) generated by new residential construction (Thomson, 2004; U.S.EPA, 2003; U.S.EPA., 1998)

Components	Content Examples	Percent (%)
Wood	Forming and framing lumber, stumps/trees, engineered wood, plywood, laminates, scraps	42.4
Drywall	Sheetrock, gypsum, plaster	27.3
Concrete and Asphalt pavement	Foundations, driveways, sidewalks, floors, road surface, sidewalks and road structures made with asphalt binder	12.0
Brick	Bricks and decorative blocks	7.3
Metals	Pipes, rebar, flashing, steel, aluminum, copper, brass, stainless steel, wiring, framing	1.8
Plastics	Vinyl siding, doors, windows, floor tile, pipes, packaging	1.4
Roofing	Asphalt & wood shingles, slate, tile, roofing felt	1.4
Glass	Windows, mirrors, lights	n.a.
Miscellaneous	Carpeting, fixtures, insulation, ceramic tile	0.6
Cardboard	From newly installed items such as appliances and tile	n.a.

n.a.: not available

2.2.3.3 Flue gas desulfurization (FGD) gypsum

Flue gas desulfurization (FGD) gypsum is a unique synthetic product derived from FGD systems at coal-based electric power plants. These systems operate by injecting absorbents such as limestone to combine with the sulfur resulting in a slurry that is mostly composed of excess lime, calcium sulfite and calcium sulfate (Karnachuk et al., 2002).

Some power plants can produce FGD gypsum which is nearly identical to mined natural gypsum. According to the American Coal Ash Association's annual Coal Combustion Product Production and Use Survey, the total production of FGD gypsum in 2006 was approximately 12 million tons. Close to 9 million tons of FGD gypsum was put to beneficial use (80% use in gypsum drywall products and 2% in agriculture), while the remainder was landfilled (U.S.EPA., 2008). However, several power plants cannot produce high purity gypsum and it becomes a solid waste instead of a commercial product. This solid by-product must then be disposed of in an approved manner. For instance, at the Mae Moh coal-fired power plant (Thailand), only 1% of the FGD gypsum can be sold, while the rest is disposed of into landfill sites due to its impurities such as fly ash and iron oxide (Panpa, 2002) (Figure 2.1d). FGD gypsum is one of the many solid waste materials which may lead to H_2S odor problems.

2.3. Sulfate Reduction in Sediments (Natural Systems)
2.3.1 River and lake sediments

The sulfate concentration in freshwater lakes and rivers is low. Therefore, the sulfur cycling has often been neglected in studies on organic matter cycling in freshwater sediments (Capone & Kiene, 1988). Most known types of SRB in freshwater systems grow best in media with low salt concentrations (maximum 0.4 g Cl^- L^{-1}), consistent with the low salinity of the freshwater habitat (Bak & Pfennig, 1991b). The optimum growth of SRB occurs in the absence of NaCl (Azabou et al., 2007a).The sulfate concentration in freshwater is about 10 to more than 500 µM, which is much lower as compared to seawater (28 mM). Oligotrophic lakes generally have a sulfate concentration below 300 µM, whereas concentrations as high as 700-800 µM have been found in meso- and eutrophic lakes (Lamers et al., 1998).

In freshwater systems, sulfate reduction rates are generally low because of the modest availability of sulfate (Lamers et al., 1998). For example, sulfate reduction rates observed in Little Rock Lake (oligotrophic lake) in northern Wisconsin were 0.48-10.8 nmol mL^{-1} d^{-1}, which were strongly influenced by temperature (Urban et al., 1994). Because of its low concentration, sulfate usually penetrates only to less than 10 cm into freshwater sediments (Cook & Schindler, 1983). Therefore, the top 10 cm of a sediment has the maximum sulfate reduction activity (Ingvorsen et al., 1981). Bacterial populations are abundant in near surface sediments, reflecting high mineralization rates and then decrease exponentially with sediment depth (Capone & Kiene, 1988; Li et al., 1996). The sulfate reduction rates were lower in the deeper (2-4 cm) than in the shallower (0-2 cm) depth intervals of Mono Lake (a hypersaline soda lake in California) sediments measured in flow-through reactors containing intact sediment slices with the incubation temperatures ranging from 10 to 50°C (Stam et al., 2010). However, a very high sulfate reduction rate of 1488 nmol mL^{-1} d^{-1} was found in the 40°C reactor. The sulfate reduction rates increased 2-5 times, with a maximum value of 4224 nmol mL^{-1} d^{-1} when lactate was added into the system.

Wellsbury et al. (1996) found that at the freshwater site Ashleworth Quay (U.K.), methanogenesis was responsible for the bulk of organic carbon mineralization (55.7%). However, sulfate reduction was still significant (13.2% of total organic carbon mineralization). Sulfate reduction rates (Thymidine incorporation measurements) decreased with depth from 52.5 nmol mL^{-1} d^{-1} in the near surface sediment to 19.8 nmol mL^{-1} d^{-1} in the 3-4 cm depth horizon, with a small increase to 48.4 nmol mL^{-1} d^{-1} at 4-5 cm sediment depth (Table 2.3).

Sulfate concentrations below 3 mM are limiting to SRB in sediments (Boudreau & Westrich, 1984; Capone & Kiene, 1988). In contrast, Ingvorsen et al. (1981) found that rates of SRB in the sediment were not sulfate limited at sulfate concentrations exceeding 0.2 mM in short-term experiments. Moreover, high sulfate reduction rates were observed at the sediment surface in Lake Mendota (eutrophic lake), Madison (USA). Sulfate reduction rates in this lake varied from 50 to 600 nmol mL^{-1} d^{-1} (measured with ^{35}S), depending on temperature and sampling date. This indicates that SRB in freshwater sediments have acquired high affinity uptake systems for sulfate in order to cope with low sulfate concentrations (Ingvorsen & Jørgensen, 1984).

Dissolved anion concentrations and sulfate reduction rates show intensive short- and long-term variations consistent with the strong seasonal changes of temperature and water level. Holmer and Storkholm (2001) concluded that sulfate reduction was predominant when the mineralization was low in winter and spring, whereas methanogenesis was most important when the overall mineralization was high in summer and autumn. In contrast, the sulfate reduction rates at the littoral site of Lake Constance (German-Swiss border) were the lowest just after the spring thaw (300-400 nmol cm^{-2} d^{-1}), but increased rapidly toward summer and reached a maximum of more than 2000 nmol cm^{-2} d^{-1} in September. Moreover, the sulfate reduction rates increased gradually from 800 nmol mL^{-1} d^{-1} at 0°C to 14250 nmol mL^{-1} d^{-1} at 40°C (Bak & Pfennig, 1991a). In Lake Kizaki (mesotrophic lake in Japan), the sulfate reduction tended to be high in spring and summer (Li et al., 1999). This is also supported by the study of David and Mitchell (1985): rates of sulfur deposition measured in sediment traps were the highest after spring turnover.

Sulfur deposition is controlled by the rate of sulfate reduction and sulfide re-oxidation (Dornblaser et al., 1994). Re-oxidation of sulfides occurs rapidly through several pathways, both under oxic and anoxic conditions. Examples of re-oxidation of sulfide are chemical oxidation with oxygen, bacterial oxidation under aerobic conditions, phototrophic oxidation, anoxic chemical oxidation and bacterial oxidation under anoxic conditions (Elsgaard & Jørgensen, 1992). High rates of re-oxidation of reduced sulfur compounds in freshwater sediments may in some cases revert the sediments from a sulfur sink to a source of sulfate to the overlying water (Bak & Pfennig, 1991a; Elsgaard & Jørgensen, 1992). Re-oxidation is high in vegetated littoral sediments because of the release of oxygen from aquatic macrophytes (Sand-Jensen et al., 1982). Bak and Pfennig (1991a) found that the total sediment sulfur at the littoral site in Lake Constance (Germany) includes: 53% present in an organically bound form, 41% as pyrite and elemental sulfur and only 6% as iron monosulfide (FeS). Moreover, deposition of sulfur is generally higher in eutrophic than in oligotrophic lakes (Holmer & Storkholm, 2001).

2.3.2 Marshes and wetlands sediments

Sulfate concentrations in freshwater wetlands are generally low, in contrast to marine wetlands. Sulfate reduction rates in anaerobic freshwater sediments are thus generally rate limited by the availability of sulfate. Figure 2.3 overviews the sulfur cycle in wetlands. In addition, large amounts of sulfate are mobilized by the oxidation of sulfide deposits by oxygen during desiccation of wetlands (Schuurkes et al., 1988) and by nitrate in aquifers through chemolithotrophic denitrification (Appelo & Postma, 1993).

Sulfate is considered to be a potential biogeochemical constraint for the development of characteristic species-rich freshwater wetlands. Sulfate reduction may lead to the accumulation of dissolved sulfide in the sediment, generating a phytotoxic effect even at low concentrations (6.8-17.3 µM of sulfide) (Smolders & Roelofs, 1993). As a result, fast growing, sulfide resistant plant species may outcompete characteristic plant species, leading to a loss of biodiversity in wetlands (Lamers et al., 1998; Smolders & Roelofs, 1993).

The response of different freshwater wetlands to sulfur pollution is, however, expected to vary because of the variation in factors controlling the sulfate reduction rates (Table 2.3). In acidic environments, sulfate reduction rates are much lower than in neutral or more alkaline environments (Roelofs, 1991). Electron donors such as acetate and lactate are other important limiting factors for the sulfate reduction process. Since these low molecular weight organic acids are the product of overall decomposition, one may predict that the easier decomposable soil organic matter is, the stronger will be its response to sulfate pollution (Lamers et al., 2001).

Table 2.3. Sulfate reduction rates in soils and sediments by radioisotope tracer experiments with ^{35}S-labeled

Soils/Sediments	Temperature (°C)	Sulfate reduction rate (nmol mL^{-1} d^{-1})	(mmol m^{-2} d^{-1})	References
Soil	20	28.8-564		Koydon (2004)
Oligotrophic Lakes				
Little Rock (Laboratory)	4-30	0.48-10.8		Urban et al. (1994)
Little Rock (Intact cores)	4-23	0-1680		Urban et al. (1994)
Constance	-10-25	300-2000		Bak and Pfennig (1991a)
Mesotrophic Lakes				
Kizaki	6	0.5-13		Li et al. (1999)
Washington	n.a.	1.73		Kuivila et al. (1989)
Kinnereret	13-30	12-1700		Hadas and Pinkas (1995)
Eutrophic Lakes				
Mendota	1-13	50-600		Ingvorsen et al. (1981)
Hypersaline Lakes				
Mono Lake	40	1488		Stam et al. (2010)
Rivers				
Ashleworth Quay	15.5	19.8-52.5		Wellsbury et al. (1996)
Colne	6-18	76.2-105.7		Kondo et al. (2007)
Estuarine				
Kingoodie Bay	14.5	58-260		Wellsbury et al. (1996)
AustWarth	16.5	58.4		Wellsbury et al. (1996)
Colne	6-18	10.2-193.4		Kondo et al. (2007)
Tomales Bay	n.a.	0-1080		Chambers et al. (1994)
Scheldt Estuary	21	240-1104		Pallud and Van Cappellen (2006)
Scheldt Estuary	30	1176		Stam et al. (2011)
Sea				
Black Sea	n.a.	0.65-1.43		Jørgensen et al. (2001)
Baltic Sea	n.a.		2.90	Thode-Andersen and Jørgensen (1989)
The Eastern Mediterranean sea	n.a.		0.1-66	Omoregie et al. (2009)

n.a.: not available

Sulfur Cycle in Wetlands

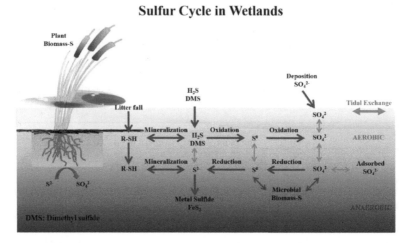

Figure 2.3. The sulfur cycle in wetlands (adapted from Inglett, 2008).

2.3.3 Mangrove sediments

Mangroves are important tropical and subtropical plant communities occurring at the interface between the land and sea (Nedwell et al., 1994). Mangroves are one of the most productive ecosystems in the world and support highly developed detritus-based food webs (Robertson, 1986). Organic detritus produced in mangrove swamps will either be degraded and recycled in the sediments or exported to adjacent areas (Kristensen et al., 1988). The large input of organic matter supports high rates of heterotrophic metabolism. Generally anaerobic conditions prevail in mangrove sediments with an overlying aerobic zone. Aerobic respiration and anaerobic sulfate reduction are usually considered the most important respiration processes in mangrove sediments (Alongi, 1998), with about 40–50% share of each. In the aerobic zone, organic matter decomposition usually proceeds by aerobic respiration. However, in the underlying anaerobic zone, decomposition occurs mainly through anaerobic processes, as sulfate reduction.

Indeed, sulfate reduction is known to be the major mineralization process in mangrove areas (Mackin & Swider, 1989). According to the study of Kristensen et al. (1991), sulfate reduction could account for almost the entire CO_2 released from a mangrove sediment in Phuket (Thailand). Most mangrove sediments consequently contain high levels of reduced inorganic sulfur in the form of primarily pyrite and elemental sulfur and only negligible amounts of iron monosulfide (Holmer et al., 1994).

The surface layer (0–10 cm) of a mangrove sediment is characterized by high bacterial numbers and high H_2S production rates (Kristensen et al., 1991). This part of the sediment appeared to be the most active, presumably due to the availability of higher amounts of organic matter. The top layer of the sediment surface contained high amounts of fallen leaves, pieces of wood, roots and mangrove fruits (Lyimo et al., 2002).The H_2S production rates in untreated sediment were 10 to 200 times higher compared to the methane (CH_4) production rates (Lyimo et al., 2002). Lyimo et al. (2002) also concluded that methanogens were outcompeted by SRB for common substrates in their mangrove sediments investigated because the inhibition of methanogens by 2-bromoethanesulfonic acid (BES) did not result into measurably higher H_2S production rates from the sediments (Lyimo et al., 2002).

Kristensen et al. (1991) also reported sulfate reduction rates higher inside a mangrove than at its periphery. These data strongly suggest that the availability of organic carbon for mineralization in the sediment increased as the transect entered the mangrove forest. In the middle of the mangrove, the proportion of organic carbon mineralized by sulfate reduction exceeded that attributable to oxygen uptake at the surface of the sediment (147%). This implied possible subsurface sources of organic electron donors for the SRB in the sediment.

2.3.4 Sea sediments

Sulfate reduction predominates especially in sediments underlying highly productive and oxygen-depleted coastal waters. Over 50% of the accumulated organic matter is mineralized in coastal and shelf sediments via sulfate reduction (Jørgensen, 1982) and it can be up to 100% of the overall organic carbon mineralization, such as in the Black Sea (Jørgensen et al., 2004). Figure 2.4 shows the dominant oxidants for mineralization changing with depth. However, the sulfate reduction rates in marine sediments tend to decrease with increasing distance from land, and it has been estimated that over 90% of oceanic sulfate reduction occurs within sediment located between the shoreline and 200 m depth (Jørgensen, 1982). Sulfate reduction rates in marine sediments have been studied by many researches (Table 2.3). Edenborn et al. (1987) found that the maximum sulfate reduction rate in sediments of four deep stations in the Saguenay Fjord, the Laurentian Trough and Gulf of Saint Lawrence (Canada), were 0.4-7.0 nmol mL^{-1} d^{-1}. In addition, the bacterial sulfate reduction rate in bottom sediments of the Gulf of Gdańsk (Baltic Sea), Poland varies from 1.89 to 31.6 nM SO_4^{2-} g^{-1} d^{-1} (Mudryk et al., 2000). Sulfate reduction rates in the Scheldt estuary sediment (The Netherlands) were determined using flow-through reactors containing intact sediment slices (Pallud & Van Cappellen, 2006; Stam et al., 2011). Sulfate reduction rates as high as 240-1104 nmol mL^{-1} d^{-1} were found at 21°C in the top 0-6 cm interval of the marine sediment (Pallud & Van Cappellen, 2006). The highest sulfate reduction rate of 1176 nmol mL^{-1} $d^{-1}$was found at 30°C in the sediment collected closest to the vegetated marshes (Stam et al., 2011).

SRB, which generate large amounts of toxic H_2S in aquatic ecosystems, are important not only for ecological reasons but they also have an economic impact. For example, in the petroleum industries, which use large amounts of seawater in their technologies while recovering oil from under the sea bed, a large amount of SRB may cause the oil and gas to acidify, the piping to corrode and technical installations to become clogged (Gibson et al., 1987; Peng et al., 1994).

Sulfate reduction with methane as electron donor occurs in marine sediments (Meulepas et al., 2010). Methane is oxidized biologically in the absence of oxygen especially at the transition between sulfate and methane. A large number of studies based on radiotracer experiments showed a maximum anaerobic oxidation of methane (AOM) and sulfate reduction rate at the methane sulfate transition zone, where sulfate and methane reach the same molar concentrations (Iversen & Jørgensen, 1985). This distinct zone is typically located one to several meters below the sediment surface in continental margin sediments and plays a key role in the biogeochemistry of the sea bed (Jørgensen & Kasten, 2006). The integrated rates of AOM in the transition zone accounted for 89% of the sulfate reduction at this depth (Iversen & Jørgensen, 1985). The AOM rate depends on a variety of conditions including the organic content of the sediments, methane supply rate, sulfate penetration in the sediments, temperature and pressure.

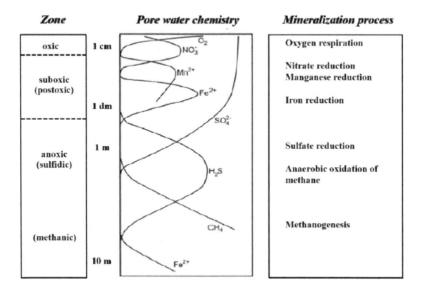

Figure 2.4. Change with depth of the dominant oxidants for mineralization in sediments (adapted from Froelich et al., 1979).

Sulfate reduction with methane as electron donor can be applied for sulfate removal and metal precipitation. However, the microorganisms involved in AOM coupled to sulfate reduction are extremely difficult to grow in vitro (Meulepas et al., 2009a). Meulepas et al. (2009a) showed that sulfate reduction with methane as electron donor is possible in well-mixed bioreactors and the submerged-membrane bioreactor system is an excellent system to enrich slow-growing microorganisms such as methanotrophic archaea. The optimum pH, salinity and temperature for SRB with methane as electron donor were 7.5, 30% and 20°C, respectively (Meulepas et al., 2009b). The volumetric AOM and sulfate reduction rates doubled approximately every 3.8 months at 15°C and the AOM and sulfate reduction rates of the obtained enrichment were 1.0 mmol g^{-1}VSS d^{-1} after 884 d of operation (Meulepas et al., 2009a; Meulepas et al., 2009b).

2.4. Biological Treatment of Sulfate Minerals
2.4.1 Biological treatment process using sulfate reducing bacteria (SRB)

Presently, a variety of physico-chemical treatment processes are employed for sulfate removal such as ion exchange, adsorption and membrane filtration. These technologies are, however, relatively expensive due to their higher operation and maintenance costs as well as energy consumption (Ozacar et al., 2008).

Biological transformation of sulfur compounds are carried out by microorganisms. The microorganisms from the sulfur cycle offer unique opportunities for sulfur pollution abatement and sulfur recovery. Sulfur compounds are an energy source in the presence of oxygen or nitrate, but they act as electron acceptor under anaerobic conditions (Lens & Kuenen, 2001). Table 2.4 and Figure 2.5 show a summary of reactions and bacteria involved in the sulfur cycle.

Table 2.4. Reaction stoichiometry and bacterial groups involved in the biological sulfur cycle

Reaction		Bacteria	References
Sulfate reduction	$SO_4^{2-} + 8e^- + 4H_2O \rightarrow S^{2-} + 8OH^-$	Sulfate reducing bacteria: *Desulfobacter* sp., *Desulfococcus* sp., and *Desulfonema* sp.	Al-Zuhair et al. (2008); Koydon (2004); Madigan et al. (2003)
Sulfide oxidation	$6CO_2 + 12H_2S \rightarrow 6(CH_2O) + 12S^0 + 6H_2O$ $2H_2S + O_2 \rightarrow 2S^0 + 2H_2O$ $H_2S + 2O_2 \rightarrow SO_4^{2-} + 2H^+$	Phototrophic bacteria: Chromatiaceae or Chlorobiaceae Chemoautotrophic bacteria: *Beggiatoa* sp. and *Thiothrix* sp.	Koydon (2004); Madigan et al. (2003)
Sulfur oxidation	$S^0 + 1.5O_2 + H_2O \rightarrow H_2SO_4$	Chemolithotrophic: *Thiobacillus* sp.	Koydon (2004); Madigan et al. (2003)

The biological sulfate reduction approach involves the use of anaerobic SRB, which reduce sulfate to sulfide by oxidizing an organic carbon source (Equation 2.2):

$$2CH_2O + SO_4^{2-} + 2H^+ \rightarrow H_2S + 2CO_2 + H_2O \qquad (2.2)$$

where CH_2O represents a simple organic compound. Biologically generated sulfide easily precipitates many of the dissolved metal ions as metallic sulfides (Gibert et al., 2004). Moreover, by mineralizing the organic substrate, the overall process results in increasing the alkalinity and pH of the wastewater. The generated carbonate and hydroxide ions may also contribute to metal removal (Dvorak et al., 1992).

Formation of biogenic sulfide is the first step for removal and recovery of sulfur or heavy metals. Sulfide can precipitate with many of the metals which may be present in wastewater (Equation 2.3):

$$H_2S + M^{2+} \rightarrow MS_{(s)} + 2H^+ \qquad (2.3)$$

where M represents metals such as iron (Fe), zinc (Zn), nickel (Ni), copper (Cu) and lead (Pb). The overall process of sulfate reduction leads to an increase in alkalinity and pH of the wastewater (Brown et al., 2002).

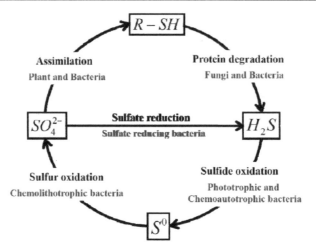

Figure 2.5. Schematic representation of the biological sulfur cycle.

2.4.2 Biological sulfate reduction

SRBs, a group of anaerobic bacteria (e.g., *Desulfovibrio, Desulfotomaculum*), have specific environmental requirements, which must be met to enable sulfate reducing activity, such as an anaerobic environment (a redox potential below -200 mV is generally needed), pH values between 5 and 8, availability of an organic substrate or hydrogen gas (H_2) to be oxidized as energy source (electron donor), availability of an appropriate sulfur species as sulfate to be reduced (electron acceptor), and a physical support on which the SRBs can be immobilized (Gibert et al., 2002).

Costa et al. (2007) found that no SRB activity was observed at pH 2. On the other hand, SRB growth was observed at pH 5 and 7 and SRB growth was not significantly different within this pH range (5 and 7). According to the study of Al-Zuhair et al. (2008), the optimum temperature and pH for mesophilic SRB were 35°C and 7, respectively. O' Flaherty et al. (1998) found that the pH optima for growth of pure cultures of SRB were between 7.5 and 8.0, whereas the pH optima of the SRB from anaerobic sludge was in the range of 7.5-8.5. This was higher than observed by pure SRB cultures. There was an increase in the maximum net specific growth rates of the SRB from pH 6.8 until their pH optima.

A carbon source and an electron donor are the primary nutrient requirements for SRB. Carbon is needed to build new bacterial cells. Possible carbon sources or electron donors include: organic acids such as formate, acetate, propionate and butyrate, various alcohols such as methanol, ethanol as well as more complex organic matter as primary sewage sludge, spent yeast from breweries, dairy whey, molasses, tannery wastewater, and micro-algal biomass. A proper carbon source and electron donor is chosen based on its cost and availability (Rzeczycka & Blaszczyk, 2005). The choice of appropriate electron donors also depends on the operational conditions as well as the species of SRB. High sulfate removal rates are achieved by using H_2/CO_2 (30 g L^{-1} d^{-1}), acetate (28.5 g L^{-1} d^{-1}), and ethanol (21 g L^{-1} d^{-1}) (de Smul et al., 1997; de Smul & Verstraete, 1999; van Houten et al., 1994).

When hydrogen is used as the electron donor, carbon monoxide (CO) or CO_2 is used as the carbon source (Moosa et al., 2002; Zhao et al., 2010). However, lactate is reported as the best suited carbon source (Koydon, 2004; Postgate, 1984) as many species of sulfate reducers can use it (Liamleam & Annachhatre, 2007). Acetate is a key intermediate in the breakdown of organic substances in anaerobic processes and can be used as an electron donor in the sulfate reduction process. However, when incompletely oxidizing sulfate reducers are present, acetate will be not utilized. Acetate production during the biological sulfate reduction is actually a major drawback of high rate sulfate reducing bioreactors because many SRB cannot completely oxidize acetate even in excess of sulfate (Lens et al., 2002).

As the prices of many simple compounds are high, residues from agriculture and wastes from the food industry become an interesting option as these waste products can be used as electron donor. Leaves, wood chips, compost, and sewage sludge have been used as electron donor for SRB. Waybrant et al. (1998) conducted batch sulfate reduction tests in order to select reactive mixtures for AMD treatment. Composted leaf mulch, composted municipal sewage sludge, maple sawdust, mixed hardwood and softwood chips, composted sheep manure and delignified waste cellulose, were tested as carbon sources. The results showed that the mixture containing sewage sludge achieved the fastest acclimation. Moreover, the sulfate reduction rate was generally higher in the reactive mixture which contained a variety of organic sources. The mixture that contained five different organic sources (sewage sludge, leaf mulch, wood chips, sheep manure and sawdust) yielded the highest sulfate reduction rate (4.23 mg L^{-1} d^{-1} g^{-1} of organic matter).

2.4.3 Solid sulfate as electron acceptor

The development of bioreactors for sulfate rich wastewater treatment, such as AMD, has been thoroughly investigated. In this case, the dissolved sulfate ion is used as electron acceptor. However, there are only a few studies focusing on the use of sulfate present in the solid phase, such as gypsum and barite ($BaSO_4$), as electron acceptor for SRB (Alfaya et al., 2009; Karnachuk et al., 2002).

In some researches, waste containing gypsum, such as phosphogypsum, was used as solid sulfate source for biological sulfate reduction (Azabou et al., 2005; Hiligsmann et al., 1996; Kowalski et al., 2003; Wolicka & Kowalski, 2006). These gypsum containing wastes are shown to be good sources of sulfate for SRB and thus sulfur and metal recovery can be achieved (Azabou et al., 2007b). However, SRB growth and activities can be inhibited due to impurities such as heavy metals present in the gypsum waste (Azabou et al., 2005; de Vries, 2006). The relative order for the inhibitory metal concentration, based on the 50% inhibitory concentration (IC_{50}) values, is Cu, Te > Cd > Fe, Co, Mn > F, Se > Ni, Al, Li > Zn (Azabou et al., 2007a).

Karnachuk et al. (2002) tested hannebachite ($CaSO_3 \cdot 0.5H_2O$), gypsum, anglesite ($PbSO_4$), and barite as electron acceptors for SRB with lactate as the electron donor. Biogenic sulfide formation occurred with all four solid phases, and protein data confirmed that bacteria grew with these electron acceptors. Sulfide formation from gypsum was almost comparable in rate and quantity to that produced from a soluble sulfate salt (Na_2SO_4). Barite as the electron acceptor supported the least growth and H_2S formation. The least soluble minerals produced the least amount of sulfide as compared to the other electron acceptors. These studies highlight the dissolution of the solid phase prior to sulfate reduction (Karnachuk et al., 2002; Kowalski et al., 2003). Moreover, the dissolution process could be accelerated by the production of extracellular polymeric substances by SRB (ZinKevich et al., 1996). However, the results demonstrate that low-solubility crystalline phases can be biologically reactive under reducing conditions (Karnachuk et al., 2002).

Gypsum has a solubility of 2600 mg L^{-1} in pure water at 25°C (FAO, 1990), which results in a sulfate concentration of 1450 mg L^{-1}. However, the sulfate concentration of the leachate can exceed the solubility limit due to the presence of other ions and the increased ionic strength of the leachate (Jang & Townsend, 2001). For instance, gypsum solubility was found to be 3 times higher in the presence of sodium chloride at 5 g L^{-1} (Shternina, 1960). Supersaturation of sulfate can also occur due to sorption of calcium by organic matter, the presence of colloidal gypsum particles and the presence of other calcium- and/or sulfate-containing mineral colloidal particles (van Den Ende, 1991).

Zegeye et al. (2007) also showed that Fe(II-III) hydroxysulfate minerals, such as green rust, support bioreduction processes by serving as electron acceptors for SRB. According to the study of Gramp et al. (2009), schwertmannite ($Fe_8O_8(OH)_6(SO_4)$), jarosite ((K, NH_4, $H_3O)Fe_3(SO_4)_2(OH)_6$), and gypsum were used as solid-phase electron acceptors for SRB with lactate as electron donor. The formation of greigite (Fe_3S_4) from schwertmannite in a sulfate reducing culture was verified with X-ray diffraction spectroscopy (Gramp et al., 2009). Greigite was also identified in solid gypsum, whereas jarosite was much less abundant. Moreover, the relative amount and crystallinity of gregite increased with the incubation temperature (Gramp et al., 2009).

A new bioremediation technology by SRB to remove the gypsum content of calcareous gypsiferous soils was investigated by Alfaya et al. (2009). Calcareous gypsiferous soils from Spain were shown to contain an endogenous SRB population which can carry out sulfate reduction using the sulfate from gypsum in the soil as electron acceptor. However, the organic matter content of this soil was rather low, so that an external electron donor (lactate) for the SRB needed to be supplied.

2.4.4 Ex situ versus in situ treatment concepts

Gypsum wastes can be treated in different ways dependent on the application. Most simple treatments (*ex situ* treatment) of these wastes are chemical or physical treatment such as washing, wet sieving, or neutralization with lime (Tayibi et al., 2009). Some of these wastes can be treated by thermal treatment to produce anhydrite for construction and cement industry applications (Manjit & Mridul, 2000; Taher, 2007).

At landfill sites (*in situ* treatment), the utilization of specific cover material to control H_2S emissions can be a useful alternative technique which is cheaper than landfill gas collection systems. Lime- and fine concrete- amended soil demonstrated the best performance in reducing H_2S emissions compared to clayey and sandy soils (Plaza et al., 2007). Plaza et al. (2007) also concluded that the particle size of the cover material is important, as the amount of sorption will increase with an increase in available surface area. However, this kind of treatment is an end-of-pipe solution, which may be insufficient.

Gypsum waste was also shown to be a good source of sulfate for SRB. Therefore, biological sulfate reduction systems can be applied for gypsum wastes treatment (Azabou et al., 2007b; Wolicka & Borkowski, 2009). There are two strategies for removal of sulfate from gypsum containing materials by biological sulfate reduction: *in situ* or *ex situ* biological sulfate reduction, which can be done in both indirect and direct treatment (Figure 2.6). For the indirect biological sulfate reduction, sulfate needs to be leached out from the gypsum waste by water and the dissolved sulfate in the leachate is subsequently removed by a biological sulfate reduction process. Nowadays this treatment concept has been studied to treat CDDS (de Vries, 2006; Kijjanapanich et al., 2013) and gypsiferous soils (Alfaya et al., 2009).

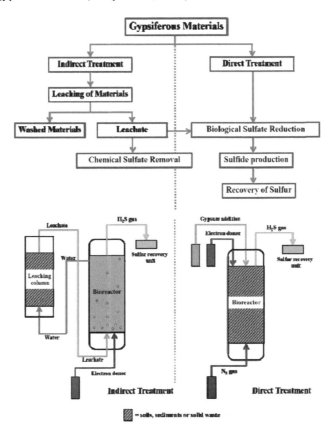

Figure 2.6. Treatment technologies for gypsiferous materials.

The direct treatment for biological sulfate reduction is another option to treat gypsum containing materials. For direct treatment, both soluble and solid sulfate in the material will be used as electron acceptor for biological sulfate reduction, and the produced sulfide needs to be trapped in a sulfide absorbing solution and treated in a further step. This type of treatment requires that anaerobic conditions are maintained at the treatment sites to ensure sulfate reduction activity. Fermentative processes might be an interesting way to achieve anaerobic field conditions (Alfaya et al., 2009) and supply of electron donor. However, studies about the *in situ* biological removal of sulfate in gypsum containing materials are rare.

2.5. Sulfate Reduction in Soils and Solid Wastes (Bioengineered Systems)
2.5.1 Soils

Research on bioremediation of gypsiferous soils especially using SRB is rare. Most of the gypsiferous soils have a relatively low organic matter content (Ghabour et al., 2008). Therefore, sufficient electron donor for SRB needs to be added when the soils are treated by biological sulfate reduction.

A novel bioremediation technology to remove the gypsum content of calcareous gypsiferous soils by SRB was investigated by Alfaya et al. (2009). Calcareous gypsiferous soils were shown to contain an endogenous SRB population that uses the sulfate from gypsum in the soil as electron acceptor. The sulfate reduction rate increased (twice faster) when anaerobic granular sludge was added to bioaugment the soil with SRB. In the presence of anaerobic granular sludge, a maximum sulfate reduction rate of 567 mg L^{-1} d^{-1} was achieved with propionate as the electron donor.

Koydon (2004) found that the population density of SRB decreased with the depth of the soil profile in as and column experiment. The SRB population was decreased slightly in the first 5 cm of the column from 7.9 $\times 10^4$ to 4.4×10^4 CFU g^{-1} dry soil. The ratio of aerobic and anaerobic bacteria changed with depth of the column. At a depth of 0-1 cm and after 30 cm, the numbers of aerobic bacteria exceeded that of the anaerobic bacteria. Aerobic conditions prevailed on top and at 30 cm depth of the sand column close to the outlet. The population densities of anaerobic bacteria were high at depths of 1-20 cm. The highest sulfate reduction rate (564 nmol mL^{-1} d^{-1}) was found in the sewage sludge layer at the top of the sand column (Table 2.3), and declined with depth in the soil profile (Koydon, 2004).

2.5.2 Solid wastes
2.5.2.1 Phosphogypsum

Phosphogypsum was shown to be a good source of sulfate for SRB in many studies (Rzeczycka & Blaszczyk, 2005; Rzeczycka et al., 2004; Wolicka & Kowalski, 2006) (Table 2.5). Sulfate and other biogenic elements present in phosphogypsum are good sources for growth of SRB if organic carbon and nitrogen were supplemented to the culture medium (Rzeczycka et al., 2001).

Hiligsmann et al. (1996) studied two stage bioreactors with immobilized SRB cells on a fixed bed (Figure 2.7a). An overall bioconversion capacity of 11 kgm^{-3} d^{-1} of gypsum (60% of the gypsum fed) and 1.2 kgm^{-3} d^{-1} of dissolved organic carbon (DOC) (45% of the DOC fed) has been achieved. The best result was obtained when cheese whey was used as carbon source. However, nitrogen gas is necessary for sulfide stripping. This sulfide was then trapped in an absorber unit. An increase in sulfide concentration (up to 556 and 416 mg L^{-1}) was found in two anaerobic cultures of mesophilic and thermophilic bacteria, respectively when biotransformation of phosphogypsum was investigated (Kowalski et al., 2003). This corresponds to a 50% reduction of the sulfate content of the phosphogypsum.

The phosphogypsum concentration also affects the growth of SRB (Azabou et al., 2005). Biogenic sulfide production was found to occur at phosphogypsum concentrations up to 40 g L^{-1} when lactate was used as electron donor. Optimal growth was obtained at 10 g L^{-1} phosphogypsum. The inhibition of SRB growth at the higher concentrations of phosphogypsum could have been caused by an accumulation of toxic levels of impurities, especially fluorine and heavy metals. Heavy metals such as zinc can inhibit the growth rate of SRB (Rzeczycka et al., 2004), depending on their speciation and concentration.

Table 2.5. Sulfate reduction rates with different sources of solid sulfate (electron acceptor) and electron donor at 30 °C

Sources of solid sulfate	Electron donor	COD/SO$_4{}^{2-}$	Max. Sulfate reduction rate (mg L^{-1} d^{-1})	References
Gypsiferous soil	Glucose	n.a.	40	Alfaya et al. (2009)
Gypsiferous soil + Anaerobic granular sluge	Glucose	n.a.	313	Alfaya et al. (2009)
	Acetate + propionate	n.a.	335	
	Propionate	n.a.	567	
	Acetate	n.a.	0	
	Lactate	n.a.	276	
	Methanol	n.a.	0	
	Ethanol	n.a.	321	
Phosphogypsum	Sodium lactate	1.40	315	Rzeczycka et al. (2004)
	Ethanol	1.40	160	
	Lactate	1.13-3.38	730	Wolicka and Kowalski (2006)
	Casein	n.a.	527	
	Ethanol	1.97	636	
	Lactose	1.35-4.04	268	
	Phenol	0.48-1.40	375	
	Lactate	1.40	310	Rzeczycka and Blaszczyk(2005)
	Ethanol	1.70	250	
	Casein	2.60	280	
	Glucose	2.20	250	
	Lactose	3.00	170	
	Acetate	1.50	170	
Construction and demolition debris sand (CDDS)	Ethanol	2.24	3800	de Vries(2006)
Flue gas desulfurization (FGD) gypsum	Sewage digest	2.50	3648	Kaufman et al. (1996)

n.a.: not available

The activity of the isolated SRB depends on the carbon source employed and the environment from which the microbial communities were isolated (Wolicka, 2008). For cultures of SRB isolated from environments contaminated with petroleum-derived compounds, the highest concentration of 838 mg HS$^-$ L^{-1} was obtained with ethanol as the sole carbon source. This corresponded to the reduction of 2365 mg SO$_4{}^{2-}$ L^{-1} and a 95% reduction of the initial phosphogypsum concentration (Wolicka & Kowalski, 2006). Assemblages of anaerobic SRB were isolated from the soil polluted by oil-derived products. The most effective assemblage was that growing in Postgate medium with lactose as the sole carbon source. A reduction of 790 mg L^{-1} sulfate (reduction of 53% of phosphogypsum introduced to the medium) was observed (Wolicka & Borkowski, 2009).

2.5.2.2 Construction and demolition debris (CDD)

Sulfate removal processes from CDD material have been developed: gypsum contained in the CDDS is leached out from the material by water and the soluble sulfate in the leachate is subsequently removed by the biological sulfate reduction process.

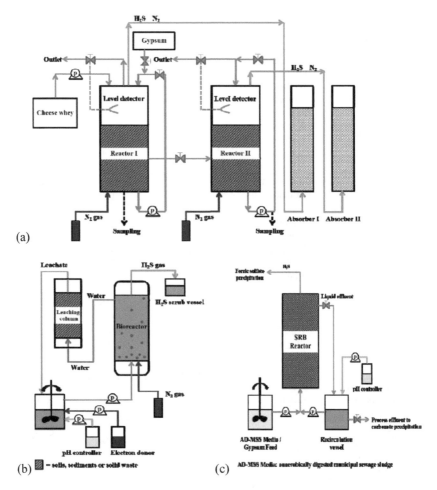

Figure 2.7. Process configuration of bioremediation technologies for: (a) phosphogypsum, (b) construction and demolition debris (CDD) and (c) flue gas desulfurization (FGD) gypsum.

To examine sulfate leachate concentrations, Jang and Townsend (2001) performed leaching experiments of CDDS. Calcium and sulfate were the predominant ions in the leachate with average sulfate concentrations ranging from 892 to 1585 mg L^{-1}. Sulfate concentrations exceeding the solubility limit (1170 mg L^{-1}) from gypsum dissolution were measured in many of the leachate samples. The mass ratio of calcium to sulfate that resulted when gypsum dissolves in solution is 0.42. Moreover, sulfate and calcium leaching had patterns similar to that of the Total Dissolved Solids (TDS), indicating that gypsum drywall from CDDS were the primary contributors to dissolved solids.

de Vries (2006) operated bioreactors for removing sulfate for the leachate of CDDS (Figure 2.7b). Sulfide formation began after a short adaptation period and about 20 g sulfate had been removed during the experiment (16 d). Ethanol dosed to the reactor was mainly used to reduce sulfate and produce acetate. The recycle flow of the other reactor was decreased by a factor 10 to give the SRB more time to reduce the sulfate. The highest sulfate reduction rate achieved was 3.8 g SO_4^{2-} $L^{-1}d^{-1}$, measured at day 16 with ethanol as electron donor (Table 2.5).

2.5.2.3 Flue gas desulfurization (FGD) gypsum

FGD gypsum can be treated in similar ways as phosphogypsum and CDD. However, research on the biotreatment of FGD gypsum, especially using SRB, is scarce and yet not well-rounded. The key factors influencing FGD gypsum treatment using SRB were investigated by Zhao et al. (2010). The experiment showed that the optimum temperature and pH for treating FGD gypsum were 37°C and 7.0-7.5, respectively. The immobilized cell reactor (Figure 2.7c), a glass column reactor filled with BIO-SEP^{TM} beads (Dupont, Glasgow, DE), was used for treating FGD gypsum (6.6 kg m^{-3} d^{-1}), with complete sulfate conversion achieved (Table 2.5) and more than 70% of the sulfur could be recovered (Kaufman et al., 1996).

2.6. Conclusions

Sulfate reduction is an important process which usually occurs in natural anaerobic environments such as in soils and sediments while discharge of sulfate to nature still does not raise much direct concern. This involves, however, other sub-processes such as iron-phosphate binding in sediments which causes eutrophication and biogenic sulfide generation causing oxygen depletion, odor and toxicity. The sulfate reduction rates depend on various factors such as the type of lakes (sulfate and organic compound concentrations), season (temperature) and depth (pressure) of the sediments.

Biological sulfate removal can be applied for treating gypsum contaminated soils, sediments and solid wastes using SRB. Recovery of elemental sulfur can also be achieved following this process. However, most of these kinds of soils, sediments and solid wastes usually have a low organic matter content. Therefore, additional organic substrates used as electron donor may need to be supplied.

2.7. References

Al-Zuhair, S., El-Naas, M. H., & Al-Hassani, H. (2008). Sulfate inhibition effect on sulfate reducing bacteria. *J. Biochem. Technol., 1*(2), 39-44.

Alfaya, F., Cuenca-Sánchez, M., Garcia-Orenes, F., & Lens, P. N. L. (2009). Endogenous and bioaugmented sulphate reduction in calcareous gypsiferous soils. *Environ. Technol., 30*(12), 1305-1312.

Alongi, D. M. (1998). *Coastal Ecosystem Processes*: CRC Press.

Appelo, C. A. J., & Postma, D. (1993). Geochemistry, Groundwater and Pollution. Rotterdam: AA Balkema.

Azabou, S., Mechichi, T., Patel, B. K. C., & Sayadi, S. (2007a). Isolation and characterization of a mesophilic heavy-metal-tolerant sulfate-reducing bacterium *Desulfomicrobium* sp. from an enrichment culture using phosphogypsum as a sulfate source. *J. Hazard. Mater., 140*, 264-270.

Azabou, S., Mechichi, T., & Sayadi, S. (2005). Sulfate reduction from phosphogypsum using a mixed culture of sulfate-reducing bacteria. *Int. Biodeter. Biodegr., 56*(4), 236-242.

Azabou, S., Mechichi, T., & Sayadi, S. (2007b). Zinc precipitation by heavy-metal tolerant sulfate-reducing bacteria enriched on phosphogypsum as a sulfate source. *Miner. Eng., 20*, 173-178.

Bak, F., & Pfennig, N. (1991a). Microbial sulfate reduction in littoral sediment of Lake Constance. *FEMS Microbiol. Ecol., 85*, 31-42.

Bak, F., & Pfennig, N. (1991b). Sulfate-reducing bacteria in littoral sediment of Lake Constance. *FEMS Microbiol. Ecol., 85*, 43-52.

Boudreau, B. P., & Westrich, J. T. (1984). The dependence of bacterial sulfate reduction on sulfate concentration in marine sediments. *Geochim. Cosmochim. Acta, 48*, 2503-2516.

Brown, M., Barley, B., & Wood, H. (2002). *Minewater Treatment*. London: IWA Publishing.

Capone, D. G., & Kiene, R. P. (1988). Comparison of microbial dynamics on marine and freshwater sediments: Contrasts in anaerobic carbon catabolism. *Limnol. Oceanogr., 33*, 725-749.

Caraco, N. F., Cole, J. J., & Likens, G. E. (1993). Sulfate control of phosphorus availability in lakes. *Hydrobiol., 253*, 275-280.

Chambers, R. M., Hollibaugh, J. T., & Vink, S. M. (1994). Sulfate reduction and sediment metabolism in Tomales Bay, California. *Biogeochem., 25*, 1-18.

Chang, I. S., Shin, P. K., & Kim, B. H. (2000). Biological treatment of acid mine drainage under sulphate-reducing conditions with solid waste materials as substrate. *Water Res., 34*(4), 1269-1227.

Cook, R. B., & Schindler, D. W. (1983). The biogeochemistry of sulfur in an experimental acidified lake. In R. Hallberg (Ed.), *Environmental Biogeochemistry* (Vol. 35, pp. 115-127). Stockholm: Ecology Bulletin.

Costa, M. C., Martins, M., Jesus, C., & Duarte, J. C. (2007). Treatment of acid mine drainage by sulfate-reducing bacteria using low cost matrices. *Water Air Soil Pollut., 189*(1-4), 149-162.

David, M. B., & Mitchell, M. J. (1985). Sulfur constituents and cycling in waters, seston, and sediments of an oligotrophic lake. *Limnol. Oceanogr., 30*(6), 1196-1207.

de Smul, A., Dries, J., Goethals, L., Grootaerd, H., & Verstraete, W. (1997). High rates of microbial sulphate redction in a mesophilic ethanol-fed expanded-granular-sludge-blanket reactors. *Appl. Microbiol. Biotechnol., 48*, 297-303.

de Smul, A., & Verstraete, W. (1999). Retention of sulfate-reducing bacteria in expaned granular-sludge-blanket reactors. *Water Environ. Res., 71*, 427-413.

de Vries, E. (2006). *Biological sulfate removal from construction and demolition debris sand*. Wageningen University.

Dornblaser, M., Giblin, A. E., Fry, B., & Peterson, B. J. (1994). Effect of sulfate concentration in the overlying water on sulfate reduction and sulfur storage in lake sediments. *Biogeochem., 24*, 129-144.

Dvorak, D. H., Hedin, R. S., Edenborn, H. M., & McIntire, P. E. (1992). Treatment of metal-contaminated water using bacterial sulfate reduction: Results from pilot-scale reactors. *Biotechnol. Bioeng., 40*(5), 609-616.

Edenborn, H. M., Silverberg, N., Mucci, A., & Sundby, B. (1987). Sulfate reduction in deep coastal merine sediments. *Mar. Chem., 21*, 329-345.

Elliott, P., Ragusa, S., & Catcheside, D. (1998). Growth of sulfate-reducing bacteria under acidic conditions in an upflow anaerobic bioreactor as a treatment system for acid mine drainage. *Water Res., 32*(12), 3724-3730.

Elsgaard, L., & Jørgensen, B. B. (1992). Anoxic transformations of radiolabeled hydrogen sulfide in marine and freshwater sediments. *Geochim. Cosmochim. Acta, 56*, 2425-2435.

FAO. (1990). *FAO Soils Bulletin 62: Management of gypsiferous soils*. Rome.

FAO. (1993). *World Soil Resources Report 66: An explanation of the FAO World Soil Resources map at a scale of 1:25 000 000*. Rome.

Froelich, P. N., Klinkhammer, G. P., Bender, M. L., Luedtke, N. A., Heath, G. R., Cullena, D., et al. (1979). Early oxidation of organic matter in pelagic sediments of the eastern equatorial Atlantic: suboxic diagenesis. *Geochim. Cosmochim. Acta, 43*(7), 1075-1090.

Ghabour, T. K., Aziz, A. M., & Rahim, I. S. (2008). Anthropogenic impact of fertilization on gypsiferous soils. *Am. Eurasian J. Agric. Environ. Sci., 4*(4), 405-409.

Gibert, O., de Pablo, J., Cortina, J. L., & Ayora, C. (2002). Treatment of acid mine drainage by sulfate-reducing bacteria using permeable reactive barrier: A review from laboratory to full-scale experiments. *Rev. Environ. Sci. Biotechnol., 1*(4), 327-333.

Gibert, O., de Pablo, J., Cortina, J. L., & Ayora, C. (2004). Chemical characterization of natural organic substrates for biological mitigation of acid mine drainage. *Water Res., 38*, 4186-4196.

Gibson, G. R., Parkes, R. J., & Herbert, R. A. (1987). Evaluation of viable counting procedures for the enumeration of sulphate-reducing bacteria estuarine sediments. *J. Microbiol. Meth., 7*, 201-210.

Gramp, J. P., Wang, H., Bigham, J. M., Jones, F. S., & Tuovinen, O. H. (2009). Biogenic synthesis and reduction of Fe(III)-hydroxysulfates. *Geomicrobiol. J., 26*, 275-280.

Gypsum Association. (1992a). Treatment and disposal of gypsum board waste: Industry position paper, *AWIC's Construction Dimensions* (Vol. March): AWIC.

Gypsum Association. (1992b). Treatment and disposal of gypsum board waste: Technical paper part II, *AWIC's Construction Dimensions* (Vol. March): AWIC.

Hadas, O., & Pinkas, R. (1995). Sulphate reduction in the hypolomnion and sediments of Lake Kinneret, Israel. *Freshw. Biol., 33*, 63-72.

Hiligsmann, S., Deswaef, S., Taillieu, X., Crine, M., Milande, N., & Thonart, P. (1996). Production of sulfur from gypsum as an industrial by-product. *Appl. Biochem. Biotechnol., 57-58*, 959-969.

Holmer, M., Kristensen, E., Banta, G., Hansen, K., Jensen, M. H., & Bussawarit, N. (1994). Biogeochemical cycling of sulfur and iron in sediments of a southeast Asian mangrove, Phuket Island, Thailand. *Biogeochem., 26*, 145-161.

Holmer, M., & Storkholm, P. (2001). Sulphate reduction and sulphur cycling in lake sediments: a review. *Freshw. Biol., 46*, 431-451.

Hulshoff Pol, L. W., Lens, P. N. L., Stams, A. J. M., & Lettinga, G. (1998). Anaerobic treatment of sulphate-rich wastewaters. *Biodegrad., 9*, 213-224.

Inglett, P. (2008). Biogeochemistry of Wetlands. Retrieved 02, 2011, from http://wetlands.ifas.ufl.edu/teaching/Biogeo-PDF-files/Lecture-10-Sulfur%20cycling-Inglett%20%5BCompatibility%20Mode%5D.pdf

Ingvorsen, K., & Jørgensen, B. B. (1984). Kinetics of sulfate uptake by freshwater and marine species of *Desulfovibrio. Arch. Microbiol., 139*, 61-66.

Ingvorsen, K., Zeikus, J. G., & Brock, T. D. (1981). Dynamics of bacterial sulfate reduction in a eutrophic lake. *Appl. Environ. Microbiol., 42*(6), 1029-1036.

Iversen, N., & Jørgensen, B. B. (1985). Anaerobic methane oxidation rates at the sulfate-methane transition in marine sediments from Kattegat and Skagerrak (Denmark). *Limnol. Oceanogr., 30*(5), 944-955.

Jang, Y. (2000). *A study of construction and demolition waste leachate from laboratory landfill simulators.* University of Florida, Florida.

Jang, Y. C., & Townsend, T. (2001). Sulfate leaching from recovered construction and demolition debris fines. *Adv. Environ. Res., 5*, 203-217.

Jong, T., & Parry, D. L. (2003). Removal of sulfate and heavy metals by sulfate-reducing bacteria in short term bench scale upflow anaerobic packed bed reactor runs. *Water Res., 37*, 3379-3389.

Jørgensen, B. B. (1982). Mineralization of organic matter in sea bed-the role of sulfate reduction. *Nat., 296*, 643-645.

Jørgensen, B. B. (1990). The sulfur cycle of freshwater sediments: Role of thiosulfate. *Limnol. Oceanogr., 35*(6), 1329-1342.

Jørgensen, B. B., Bottcher, M. E., Luschen, H., Neretin, L., & Volkov, I. (2004). Anaerobic methane oxidation and a deep H_2S sink generate isotopically heavy sulfides in Black Sea sediments. *Geochim. Cosmochim. Acta, 68*, 2095-2118.

Jørgensen, B. B., & Kasten, S. (2006). Sulfur cycling and methane oxidation. In H. D. Schulz & M. Zabel (Eds.), *Marine Geochemistry* (2nd ed.). Berlin: Springer-Verlag.

Jørgensen, B. B., Weber, A., & Zopfi, J. (2001). Sulfate reduction and anaerobic methane oxidation in Black Sea sediments. *Deep-Sea Res., 48*, 2097-2120.

Karnachuk, O. V., Kurochkina, S. Y., & Tuovinen, O. H. (2002). Growth of sulfate-reducing bacteria with solid-phase electron acceptors. *Appl. Microbiol. Biotechnol., 58*, 482-486.

Kaufman, E. N., Little, M. H., & Selvaraj, P. T. (1996). Recycling of FGD gypsum to calcium carbonate and elemental sulfur using mixed sulfate-reducing bacteria with sewage digest as a carbon source. *J. Chem. Technol. Biotechnol., 66*, 365-374.

Khresat, S. A., Rawajfih, Z., Buck, B., & Monger, H. C. (2004). Geomorphic features and soil formation of arid lands in Northeastern Jordan. *Arch. Agron. Soil Sci., 50*, 607-615.

Kijjanapanich, P., Annachhatre, A. P., Esposito, G., van Hullebusch, E. D., & Lens, P. N. L. (2013). Biological sulfate removal from gypsum contaminated construction and demolition debris. *J. Environ. Manage., 131*, 82-91.

Kijjanapanich, P., Pakdeerattanamint, K., Lens, P. N. L., & Annachhatre, A. P. (2012). Organic substrates as electron donors in permeable reactive barriers for removal of heavy metals from acid mine drainage. *Environ. Technol., 33*(23), 2635-2644.

Kondo, R., Purdy, K. J., Silva, S. Q., & Nedwell, D. B. (2007). Spatial Dynamics of sulphate-reducing bacteria compositions in sediment along a salinity gradient in a UK estuary. *Microbes Environ., 22*(1), 11-19.

Kordlaghari, M. P., & Rowell, D. L. (2006). The role of gypsum in the reactions of phosphate with soils. *Geoderma, 132*, 105-115.

Kowalski, W., Holub, W., Wolicka, D., Przytocka-Jusiak, M., & Blaszczyk, M. (2003). Sulphur balance in anaerobic cultures of microorganisms in medium with phosphogypsum and sodium lactate. *Archiwum Mineralogiczne*, 33-40.

Koydon, S. (2004). *Contribution of sulfate-reducing bacteria in soil to degradation and retention of COD and sulfate.* Karlsruhe University, Germany.

Kristensen, E., Andersen, F. B., & Kofoed, L. H. (1988). Preliminary assessment of benthic community metabolism in a south-east Asian mangrove swam. *Mar. Ecol. Prog. Ser., 48*, 137-145.

Kristensen, E., Holmer, M., & Bussarawit, N. (1991). Benthic metabolism and sulfate reduction in a southeast Asian mangrove swamp. *Mar. Ecol. Prog. Ser., 73*, 93-103.

Kuivila, K. M., Murray, J. W., & Devol, A. H. (1989). Methane production, sulfate reduction and competition for substrates in the sediments of Lake Washington. *Geochim. Cosmochim. Acta, 53*, 409-416.

Lamers, L. P. M., Dolle, G. E. T., Van Den Berg, S. T. G., Van Delft, S. P. J., & Roelofs, J. G. M. (2001). Differential responses of freshwater wetland soils to sulphate pollution. *Biogeochem., 55*, 87-102.

Lamers, L. P. M., Tomassen, H. B. M., & Roelofs, J. G. M. (1998). Sulfate-induced eutrophication and phytotoxicity in freshwater wetlands. *Environ. Sci. Technol., 32*, 199-205.

Lens, P. N. L., & Kuenen, J. G. (2001). The biological sulfur cycle: novel opportunities for environmental biotechnology. *Water Sci. Technol., 44*(8), 57-66.

Lens, P. N. L., Vallero, M., Esposito, G., & Zandvoort, M. (2002). Perspectives of sulfate reducing bioreactor in environmental biotechnology. *Rev. Environ. Sci. Biotechnol., 1*, 311-325.

Lens, P. N. L., Visser, A., Janssen, A. J. H., Hulshoff Pol, L. W., & Lettinga, G. (1998). Biotechnological treatment of sulfate rich wastewaters. *Crit. Rev. Environ. Sci. Technol., 28*(1), 41-88.

Li, J., Purdy, K. J., Takii, S., & Hayashi, H. (1999). Seasonal chages in ribosomal RNA of sulfate-reducing bacteria and sulfate reducing activity in a freshwater lake sediment. *FEMS Microbiol. Ecol., 28*, 31-39.

Li, J., Takii, S., Kotakemori, R., & Hayashi, H. (1996). Sulfate reduction in profundal sediment in Lake Kizaki, Japan. *Hydrobiol., 333*, 201-208.

Liamleam, W. (2007). *Zinc removal from industrial discharge using thermophilic biological sulfate reduction with molasses as electron donor.* Asian Institute of Technology, Thailand.

Liamleam, W., & Annachhatre, A. P. (2007). Electron donors for biological sulfate reduction. *Biotechnol. Adv., 25*(5), 452-463.

Lloyd, G. M. (1985). *Phosphogypsum: A review of Florida institute of phosphate research programs to develop uses for phosphogypsum.* Florida: Florida Institute of Phosphate Research.

Lyimo, T. J., Pol, A., & Op den Camp, H. J. M. (2002). Sulfate reduction and methanogenesis in dediments of Mtoni mangrove forest, Tanzania. *J. Hum. Environ., 31*(7), 614-616.

Mackin, J. E., & Swider, K. T. (1989). Organic matter decomposition pathways and oxygen consumption in coastal marine sediments. *J. Mar. Res., 47*, 681-716.

Madigan, M. T., Martinki, J. M., & Parker, J. (2003). *Brock biology of microorganisms* (Internation ed.). USA: Prentice Hall.

Manjit, S., & Mridul, G. (2000). Making of anhydrite cement from waste gypsum. *Cement and Concrete Research, 32*(7), 1033-1038.

Mays, D. A., & Mortvedt, J. J. (1986). Crop response to soil applications of phosphogypsum. *J. Environ. Qual., 15*, 78-81.

Meulepas, R. J. W., Jagersma, C. G., Gieteling, J., Buisman, C. J. N., Stams, A. J. M., & Lens, P. N. L. (2009a). Enrichment of anaerobic methanotrophs in sulfate-reducing membrane bioreactors. *Biotechnol. Bioeng., 104*(3), 458-470.

Meulepas, R. J. W., Jagersma, C. G., Khadem, A. F., Buisman, C. J. N., Stams, A. J. M., & Lens, P. N. L. (2009b). Effect of environmental conditions on sulfate reduction with methane as electron donor by an Eckernforde Bay enrichment. *Environ. Sci. Technol., 43*, 6553-6559.

Meulepas, R. J. W., Stams, A. J. M., & Lens, P. (2010). Biotechnological aspects of sulfate reduction with methane as electron donor. *Rev. Environ. Sci. Biotechnol., 9*, 59-78.

Montero, A., Tojo, Y., Matsuto, T., Yamada, M., Asakura, H., & Ono, Y. (2010). Gypsum and organic matter distribution in a mixed construction and demolition waste sorting process and their possible removal from outputs. *J. Hazard. Mater., 175*, 747-753.

Moosa, S., Nemati, M., & Harrison, S. T. L. (2002). A kinetic study on anaerobic reduction of sulphate, part I: Effect of sulphate concentration. *Chem. Eng. Sci., 57*(14), 2773-2780.

Mudryk, Z. J., Podgorska, B., Ameryk, A., & Bolalek, J. (2000). The occurrence and activity of sulphate-reducing bacteria in the bottom sediments of the Gulf of Gdansk. *Oceanol., 42*(1), 105-117.

Nedwell, D. B., Blackburn, T. H., & Wiebe, W. J. (1994). Dynamic mature of the turnover of organic carbon, nitrogen and sulphur in the sediments of a Jamaican mangrove forest. *Mar. Ecol. Prog. Ser., 110*, 223-231.

O' Flaherty, V., Mahony, T., O'Kennedy, R., & Colleran, E. (1998). Effect of pH on growth kinetics and sulphide toxicity thresholds of a range of methanogenic, syntrophic and sulphate-reducing bacteria. *Process Biochem., 33*(5), 555-569.

Omoregie, E. O., Niemann, H., Mastalerz, V., Lange, G. J., Stadnitskaia, A., Mascle, J., et al. (2009). Microbial methane oxidation and sulfate reduction at cold seeps of the deep Eastern Mediterranean Sea. *Mar. Geol., 261*(1-4), 114-127.

Ozacar, M., Sengil, I. A., & Turkmenler, H. (2008). Equilibrium and kinetic data, and adsorption mechanism for adsorption of lead onto valonia tannin resin. *Chem. Eng. J., 143*, 32-42.

Pallud, C., & Van Cappellen, P. (2006). Kinetics of microbial sulfate reduction in estuarine sediments. *Geochim. Cosmochim. Acta, 70*, 1148-1162.

Panpa, W. (2002). *Production of dental stone from flue-gas gypsum.* Chulalongkorn University.

Peiffer, S. (1998). Geochemical and microbial processes in sediments and at the sediment-water interface of acidic mining lakes. *Water Air Soil Pollut., 108*, 227-229.

Peng, C. G., Park, J. K., & Patenaude, R. W. (1994). Statistics-based classification of microbially influenced corrosion in freshwater systems. *Water Res., 28*, 951-959.

Plaza, C., Xu, Q., Townsend, T., Bitton, G., & Booth, M. (2007). Evaluation of alternative landfill cover soils for attenuating hydrogen sulfide from construction and demolition (C&D) debris landfills. *J. Environ. Manage., 84*, 314-322.

Porta, J., & Herrero, J. (1990). Micromorphology and genesis of soils enriched with gypsum. In L. A. Douglas (Ed.), *Soil Micromorphology: A Basic and Applied Science* (pp. 321-339). Amsterdam: Elsevier.

Postgate. (1984). *The sulphate-reducing bacteria* (2nd ed.). Cambridge: Cambridge University Press.

Robertson, A. I. (1986). Leaf-burying crabs: their influence on energy flow and export from mixed mangrove forests (*Rhizophora* spp.) in northeastern Australia. *J. Exp. Mar. Biol. Ecol., 102*, 237-248.

Roelofs, J. G. M. (1991). Inlet of alkaline river water into peaty lowlands: effects on water quality and on *Stratiotes aloides* stands. *Aquat. Bot., 39*, 267–293.

Rzeczycka, M., & Blaszczyk, M. (2005). Growth and activity of sulphate-reducing bacteria in media containing phosphogypsum and different sources of carbon. *Pol. J. Environ. Stud., 14*(6), 891-895.

Rzeczycka, M., Mycielski, R., Kowalski, W., & Galazka, M. (2001). Biotransformation of phosphogypsum in media containing defferent forms of nitrogen. *Acta Mocrobiol. Pol., 50*, 3-4.

Rzeczycka, M., Suszek, A., & Blaszczyk, M. (2004). Biotransformation of phosphogypsum by sulphate-reducing bacteria in media containing different zinc salts. *Pol. J. Environ. Stud., 13*(2), 209-217.

Sand-Jensen, K., Prahl, C., & Stokholm, H. (1982). Oxygen release from roots of submerged aquatic macrophytes. *Oikos, 38*, 349-354.

Schuurkes, J. A. A. R., Kempers, A. J., & Kok, C. J. (1988). Aspects of biochemical sulphur conversions in sediments of a shallow soft water lake. *J. Freshwater Ecol., 4*, 369-381.

Shternina, E. B. (1960). Solubility of gypsum in aqueous solutions of salts. *Int. Geol. Rev., 1*, 605-616.

Smith, R., & Robertson, V. S. (1962). Soil irrigation classification of shallow soils overlying gypsum beds, northern Iraq. *J. Soil Sci., 13*, 106-115.

Smolders, A., & Roelofs, J. G. M. (1993). Sulphate-mediated iron limitation and eutrophication in aquatic ecosystems. *Aquat. Bot., 46*, 247-253.

Stam, M. C., Mason, P. R. D., Laverman, A. M., Pallud, C., & Van Cappellen, P. (2011). $^{34}S/^{32}S$ fractionation by sulfate-reducing microbial communities in estuarine sediments. *Geochim. Cosmochim. Acta, 75*, 3903-3914.

Stam, M. C., Mason, P. R. D., Pallud, C., & Van Cappellen, P. (2010). Sulfate reducing activity and sulfur isotope fractionation by natural microbial communities in sediments of a hypersaline soda lake (Mono Lake, California). *Chem. Geol., 278*, 23-30.

Taha, R., & Seals, R. K. (1992). Engineering properties and potential uses of by-product phosphogypsum *Proceedings of utilization of waste materials in civil engineering construction*. New York: American Society of Civil Engineers.

Taher, M. A. (2007). Influence of thermally treated phosphogypsum on the properties of Portland slag cement. *Resour. Conservation Recycling, 52*(1), 28-38.

Tayibi, H., Choura, M., Lopez, F. A., Alguacil, F. J., & Lopez-Delgado, A. (2009). Environmental impact and management of phosphogypsum. *J. Environ. Manage., 90*, 2377-2386.

Thode-Andersen, S., & Jørgensen, B. B. (1989). Sulfate reduction and the formation of ^{35}S-labeled FeS, FeS_2, and S^0 in coastal marine sediments. *Limnol. Oceanogr., 34*, 793-806.

Thomson, L. (2004). construction and demolition debris recovery program. Retrieved 06, 2011, from http://www.cccounty.us/depart/cd/recycle/debris.htm

Turley, W. (1998). What's happening in gypsum recycling. *C&D Debris Recycling, 5*(1), 8-12.

U.S.EPA. (2003). *Estimating 2003 building-related construction and demolition materials amounts.*

U.S.EPA. (1998). *Characterization of building-related construction and demolition debris in the United States.* Washington D.C.

U.S.EPA. (2008). Agricultural uses for flue gas desulfulfurization (FGD) Gypsum.

U.S.EPA. (2010). Radiation Protection, U.S. Environmental Protection Agency. *About phosphogypsum* Retrieved January 16, 2011, from Website: http://www.epa.gov/rpdweb00/neshaps/subpartr/about.html

Urban, N. R., Brezonik, P. L., Baker, L. A., & Sherman, L. A. (1994). Sulfate reduction and diffusion in sediments of Litter Rock Lake, Wisconsin. *Limnol. Oceanogr., 39*(4), 797-815.

van Den Ende, J. (1991). Supersaturation of soil solutions with respect to gypsum. *Plant and Soil, 133*, 65-74.

van Houten, R. T., Hulshoff Pol, L. W., & Lettinga, G. (1994). Biological sulphate reduction using gas-lift reactors fed with hydrogen and carbon dioxide as energy and carbon source. *Biotechnol. Bioeng., 44*, 586-594.

Vincke, E., Boon, N., & Verstraete, W. (2001). Analysis of the microbial communities on corroded concrete sewer pipes - a case study. *Appl. Microbiol. Biotechnol., 57*, 776-785.

Waybrant, K. R., Blowes, D. W., & Ptacek, C. J. (1998). Selection of Reactive Mixtures for Use in Permeable Reactive Walls for Treatment of Mine Drainage. *Environ. Sci. Technol., 32*(13), 1972-1979.

Wellsbury, P., Herbert, R. A., & Parkes, R. J. (1996). Bacterial activity and production in near-surface estuarine and freshwater sediments. *FEMS Microbiol. Ecol., 19*, 203-214.

Wissa, A. E. Z. (2003). Phosphogypsum disposal and the environment. Retrieved 05, 2011, from http://www.fipr.state.fl.us/pondwatercd/phosphogypsum_disposal.htm

Wolicka, D. (2008). Biotransformation of phosphogypsum in wastewaters from the dairy industry. *Bioresour. Technol., 99*, 5666-5672.

Wolicka, D., & Borkowski, A. (2009). Phosphogypsum biotransformation in cultures of sulphate reducing bacteria in whey. *Int. Biodeter. Biodegr., 63*, 322-327.

Wolicka, D., & Kowalski, W. (2006). Biotransformation of phosphogypsum in petroleum-refining wastewaters. *Pol. J. Environ. Stud., 15*(2), 355-360.

Zegeye, A., Huguet, L., Abdelmoula, M., Carteret, C., Mullet, M., & Jorand, F. (2007). Biogenic hydroxysulfate green rust, a potential electron acceptor for SRB activity. *Geochim. Cosmochim. Acta, 71*(22), 5450-5462.

Zhao, Y., Chen, C., & Han, Y. (2010). Study on treating desulfurization gypsum by sulfate-reducing bacteria. *J. Environ. Technol. Eng., 3*(1), 5-10.

ZinKevich, V., Bogdarina, I., Kang, H., Hill, M. A. W., Tapper, R., & Beech, I. B. (1996). Characterization of expolymers produced by different isolates of marine sulphate-reducing bacteria. *Int. Biodeter. Biodegr., 37*, 163-173.

CHAPTER 3

Organic Substrates as Electron Donors in Permeable Reactive Barriers for Removal of Heavy Metals from Acid Mine Drainage

This chapter has been published as:
Kijjanapanich, P., Pakdeerattanamint, K., Lens, P. N. L., & Annachhatre, A. P. (2012). Organic substrates as electron donors in permeable reactive barriers for removal of heavy metals from acid mine drainage. *Environ. Technol., 33*(23), 2635-2644. DOI: 10.1080/09593330.2012.673013

Chapter 3

This research was conducted to select suitable natural organic substrates as potential carbon sources for use as electron donors for biological sulfate reduction in permeable reactive barriers (PRB). A number of organic substrates were assessed through batch and continuous column experiments under anaerobic conditions with acid mine drainage (AMD) obtained from an abandoned lignite coal mine. To keep the heavy metal concentration at a constant level, the AMD was supplemented with heavy metals whenever necessary. Under anaerobic conditions, sulfate reducing bacteria (SRB) converted sulfate into sulfide using the organic substrates as electron donors. Sulfide that was generated precipitated heavy metals as metal sulfides. Organic substrates, which yielded the highest sulfate reduction in batch tests, were selected for continuous column experiments which lasted over 200 d. A mixture of pig farm wastewater treatment sludge, rice husk and coconut husk chips yielded the best heavy metal (Fe, Cu, Zn and Mn) removal efficiencies of over 90%.

3.1. Introduction

AMD is produced when pyrite containing mine tailings are exposed to oxygen in the atmosphere and water as per the following equations (Stumm & Morgan, 1981):

$$FeS_{2(S)} + \tfrac{7}{2}O_2 + H_2O \rightarrow Fe^{2+} + 2SO_4^{2-} + 2H^+ \tag{3.1}$$

$$Fe^{2+} + \tfrac{1}{4}O_2 + H^+ \rightarrow Fe^{3+} + \tfrac{1}{2}H_2O \tag{3.2}$$

$$FeS_{2(S)} + 14\,Fe^{3+} + 8H_2O \rightarrow 15\,Fe^{2+} + 2SO_4^{2-} + 16H^+ \tag{3.3}$$

AMD, which has a pH of 4.0 - 4.5 or lower, solubilizes heavy metals present in the mine tailings (Chang et al., 2000; Christensen et al., 1996). Due to its highly toxic nature, AMD poses a significant environmental threat. Virtually no life can survive in such acidified waters. AMD generated from abandoned mines and mine tailings have created large lagoons worldwide. Heavy metals in soluble form affect the food chain through bio-accumulation and bio-magnification, posing a greater threat to all forms of life (Gray, 1997). AMD from these lagoons percolates through soil, thereby affecting the soil chemistry and contaminating the groundwater (Gibert et al., 2011), which is a valuable source for drinking water and for agriculture.

Remediation techniques such as physico-chemical treatment by pH adjustment to the alkaline range followed by metal hydroxide precipitation have been employed (Huttagosol & Kijjanapanich, 2008; Morrison & Spangler, 1992; Morrison & Spangler, 1993; Ngwenya et al., 2006). These methods are expensive and produce large volumes of inorganic sludge which is often difficult to dispose of due to its toxicity (Elliott et al., 1998). Pump and treat remediation methods are often difficult to employ when dealing with groundwater contamination from AMD (Keely, 1989; National-Research-Council, 1994). Metal hydroxides can resolubilize the metals depending on the redox potential and pH (Masscheleyn et al., 1991). Passive treatment methods such as the PRB technology may be more appropriate (Lapointe et al., 2006; Walton-Day, 2003). PRBs can be both abiotic and biotic treatment systems (Pagnaneli et al., 2009). In the abiotic treatment system, neutralizing agents (lime), adsorbents (silica sand or perlite) or zerovalent iron are used as reactive materials (Pagnaneli et al., 2009). In biological system, PRBs employ natural organic substrates as electron donors to facilitate the growth of SRB. When an AMD plume containing sulfate and heavy metals passes through the reactive barrier, SRB in the reactive barrier convert the sulfate into sulfide while consuming the organic substrates as electron donors (Tsukamoto et al., 2004). Heavy metals present in the contaminated feed water are then removed as metal sulfides (Dvorak et al., 1992; Jong & Parry, 2003).

SRBs, which are heterotrophic by nature, require specific environmental conditions for their growth and activity such as anaerobic conditions, pH between 5-8, temperature between 20-35°C and the presence of a carbon compound which acts as nutrient and electron donor (Gibert et al., 2002). A physical support for bacterial attachment increases their concentration. However, in sub-surface soil environments, lack of readily available organic carbon is the most common limitation to biological sulfate reduction (Gibert et al., 2002).

PRBs may be designed based on the results from feasibility experiments aimed at selection of viable organic substrates. These results can be obtained through batch and column experiments conducted in the laboratory. Many types of organic substrates, such as composted municipal sewage sludge, wood chips, sheep manures, and oak leaf, were tested as electron donors for SRB (Gibert et al., 2002; Gibert et al., 2011; Pagnaneli et al., 2009; Waybrant et al., 1998). The use of natural organic substrates as electron donors for SRB in PRB is more appropriate due to their ease of availability and cost considerations (Costa et al., 2007). This research describes the results obtained from batch and continuous column experiments testing no or low cost organic substrates as electron donors for the SRB.

3.2. Material and Methods
3.2.1 Acid mine drainage (AMD)

AMD was collected from an acidified lagoon generated from leachate of an abandoned coal mine in Lamphun Province (Northern Thailand). AMD was stored in a cold room maintained at 4°C. AMD was characterized for its pH, metals content and sulfate concentration (Table 3.1). This original AMD was supplemented further by metals whenever necessary and then used as feed for the experiments (Table 3.1).

Table 3.1. Characteristics of acid mine drainage (AMD) used in batch and column tests

Parameters	AMD (batch tests)	Added	AMD (column tests)
pH	4.16±0.08	-	4.16±0.08
Sulfate, mg L^{-1}	731±55.2	-	838±65.0
Iron, mg L^{-1}	0.08±0.05	30	26.9±0.78
Manganese, mg L^{-1}	16.7±0.91	-	16.7±0.56
Copper, mg L^{-1}	0.04±0.01	20	17.5±0.53
Zinc, mg L^{-1}	0.92±0.11	5	6.35±0.03

3.2.2 Sulfate reducing bacteria (SRB) inoculums

Sludge from a full scale mesophilic anaerobic baffled reactor treating tapioca starch wastewater was used as source for SRB. The seed sludge was analyzed for its total suspended solids (TSS) and volatile suspended solids (VSS) content.

3.2.3 Organic substrates

Five organic substrates were selected for their possible use as electron donors for SRB in the PRBs. These included bamboo chips (BC), rice husk (RH), pig farm wastewater treatment sludge (PWTS), municipal wastewater treatment sludge (MWTS) and coconut husk chips (CHC), based on their availability, ease of handling and no or low cost. Organic substrates were washed by tap water, air dried, cut to the desired size and analyzed for their physical characteristics (Table 3.2).

The elemental composition of organic substrates was analyzed using a Perkin 2400 series 2 CHNS elemental analyzer after drying and grinding the samples to fine powder. The metal composition of the organic substrates was analyzed using the wet digestion method. In this method, 1 g of organic substrate was added with 10 mL concentrated nitric acid (HNO_3) and heated at 120-140°C until no change in color of organic substrate was observed. The supernatant was then filtered and analyzed for metals (Zheljazkov & Nielsen, 1996). Leaching tests of each organic substrate were carried out to determine the leachable metals. For this, 2 g of organic substrate were supplied to 50 mL of deionized water and placed in a 55 mL centrifuge tube and put on a rotary shaker for 66 h at 150 rpm. The supernatant was then filtered and analyzed for metals, sulfate and dissolved organic carbon (DOC) content.

Table 3.2. Physical characteristics of the organic substrates used in batch and column tests

Organic material	Size (cm)	Density (g dry weight cm^{-3})	Moisture content (%)	Lignin Content (%)
RH	1.0 – 1.5	0.646	7.78	24.4 (Blasi et al., 1999)
CHC	2.0 – 3.0	0.122	39.21	46.5 (Bilba et al., 2007)
BC	2.0 – 3.0	0.785	5.78	25.8 (Vu et al., 2003)
PWTS	2.0 – 3.0	0.949	22.00	Low
MWTS	1.0 – 2.0	0.624	49.78	Low

3.2.4 Batch experiments

Five organic substrates were evaluated individually in 1.5 L batch containers (Figure 3.1a) at ambient temperature (30 ± 5°C) and anaerobic conditions to assess their ability for promoting biological sulfate reduction. During the acclimatization period, biological sulfate reduction by SRB may progress at a considerably slower pace, leading to a lower alkalinity generation. Therefore, to compensate the lower alkalinity production during this period, it was necessary to adjust the pH of the AMD to the optimum range for SRB (pH 6-7). Each reaction bottle contained 20% by volume (300 mL) of each organic material, deoxygenated AMD from the abandoned lignite coal mine 66% by volume

(1000 mL), and SRB inoculum 7% by volume (100 mL). The remaining volume (7%) is the headspace of the batch bottle. Based on the results from the single substrate batch tests, 3 organic substrates were selected for mixed substrate batch tests.

(a)   (b)

Figure 3.1. Reactors of the experiment (a) the reaction bottle for the batch experiment and (b) the column reactor for the continuous experiment.

3.2.5 Continuous column experiments

Long term continuous column experiments were conducted at room temperature (30 ± 5°C) with mixtures of 3 organic substrates. Fast degrading (PWTS), moderately degrading (RH) and slow degrading (CHC) organic substrates were selected and mixed in 4 different proportions as given in Table 3.3 and filled in 4 column reactors made of polyvinyl chloride (PVC) each with a volume of 12 L (Figure 3.1b). 1.8 g of lime was mixed with organic material in each reactor prior to filling into the reactor columns (0.15 g of lime L^{-1} of AMD (Huttagosol & Kijjanapanich, 2008)).

The hydraulic retention time (HRT) required to achieve 90% removal in continuous column experiments was estimated using the following equation (Levenspiel, 1999):

$$HRT = \frac{\ln \dfrac{S_i}{S_0}}{-k} = \frac{\ln \dfrac{10}{100}}{-k}$$

(3.4)

where: S_i = effluent concentration
S_o = influent concentration
k = first order rate constant
t = time

For 90% sulfate removal the k value amounted to 0.206 d^{-1} (obtained from the mixture of PWTS, CHC and RH), and the HRT thus equals 11.2 d. Accordingly, incorporating an appropriate safety factor, a HRT of 16 d was maintained in the continuous column experiments.

Deoxygenated AMD supplemented with metals (Table 3.1) was fed at the bottom of the reactor at a flow rate of 30 mL h^{-1} (37.18 L m^{-2} d^{-1}) using a peristaltic pump. AMD travelled through the fixed bed of the organic substrate mixture which also acted as support for immobilization of SRB, while the effluent was withdrawn from the top of reactor.

Table 3.3. Ratio of each organic material used in column tests

Reactor Number	Volume (L), (% *v/v*)			Total Volume (L)
	PWTS	RH	CHC	
1	1.0 (33.3)	1.0 (33.3)	1.0 (33.3)	3
2	1.8 (60)	0.6 (20)	0.6 (20)	3
3	0.6 (20)	1.8 (60)	0.6 (20)	3
4	0.6 (20)	0.6 (20)	1.8 (60)	3

3.2.6 Analytical methods

Batch bottles and columns were periodically sampled for pH, oxidation-reduction potential (ORP), sulfate, DOC and heavy metals. pH was measured using a Mettler Toledo pH meter, while ORP was measured using a Hach ORP meter.

Alkalinity in the column experiments was measured using the titration method. DOC which was monitored as an indicator of dissolved carbon available for bacterial activity was measured using the high temperature combustion method by a Shimadzu TOC analyzer (Eaton et al., 2005).

Sulfate removal was used as an indicator of SRB activity. Sulfate was measured using the turbidimetric method by a Shimadzu UV visible spectrophotometer. Metals (Fe, Cu, Zn, and Mn) were measured using Perkin Elmer Inductive Coupled Plasma (ICP) optical spectrophotometry. All analyses were performed according to Standard Methods for examination of water and wastewater (Eaton et al., 2005). All samples were filtered using 0.45 micron GFG glass fiber filter paper for determination of alkalinity, DOC, heavy metal and sulfate. During the sampling process, care was taken to minimize sample aeration and air infiltration into the batch bottles and columns.

3.3. Results and Discussion
3.3.1 Characteristics of AMD and SRB inoculums
3.3.1.1 AMD

The AMD had a pH of 4.2 and sulfate and Mn concentrations of 731 and 16.7 mg L^{-1}, respectively. The Mn concentration exceeds the groundwater as well as the surface water quality standards of Thailand. On the other hand, Cu, Pb and Zn concentrations were below the groundwater and surface water quality standard values. Raw AMD was used in batch tests, while AMD supplemented with Fe, Cu and Zn was used in the continuous column experiments (Table 3.1).

3.3.1.2 SRB inoculums

The SRB seed sludge had a TSS of 9.78 g L^{-1} and VSS of 8.12 g L^{-1}. The VSS/TSS ratio was 0.83.

3.3.2 Batch experiments
3.3.2.1 Single substrate batch tests

Degradability of organic substrate: Table 3.2 gives the physical characteristics of the organic substrates. Since recharging of a PRB by organic substrate cannot be done frequently under field conditions, it is important that the selected organic substrate should degrade gradually and last long. Degradability of each material is defined by volatile solids to total solids ratio (VS/TS). Table 3.4 shows the VS/TS ratio at the start of the experiment and after 16 d. As the data reveal, although CHC and BC initially had a very high VS/TS ratio, it virtually remained constant after 16 d, suggesting that these substrates would degrade slowly under subsurface conditions and hence would last long. On the other hand, the VS/TS ratio for the other 3 substrates, namely RH, MWTS and PWTS showed gradual reduction during the 16 d of the experiment. MWTS and PWTS degraded faster than RH.

pH and alkalinity: The pH of the original AMD was first adjusted between 6 and 7 and then used for batch experiments. As shown in Figure 3.2a, the pH remained between 5.5-7.5 throughout the experiments for all organic substrates except BC for which the pH decreased dramatically from an initial value of 6.3 to about 4.0 within the first 2 d and then remained in the range of 4.0-4.5 thereafter.

Since the initial pH was adjusted to 6-7, the alkalinity of the AMD increased to about 9.2 mg L^{-1} as $CaCO_3$ (Figure 3.2b). Alkalinity in all batch tests increased with time, PWTS had the highest alkalinity of 4150 mg L^{-1} as $CaCO_3$ after 16 d of the experiment. RH, MWTS, CHC and BC finally reached the alkalinity of 1327, 612, 159 and 0 mg L^{-1} as $CaCO_3$, respectively.

Due to the acidic nature of AMD, it is important that the organic substrate must be capable of generating alkalinity during the progress of the sulfate reduction reaction as follows (Sawyer et al., 2003):

$$SO_4^{2-} + Organic \quad Matter \rightarrow HS^- + HCO_3^- \qquad (3.5)$$

Alkalinity generated would be utilized for neutralization of acidity from AMD. Results show that PWTS produced the maximum alkalinity, while RH produced moderate alkalinity (Figure 3.2b). MWTS and CHC produced lower alkalinity, while BC produced no alkalinity at all. These observations also show that the organic contents from PWTS are readily available for SRBs as electron donor which in turn produced alkalinity as by product of sulfate reduction. RH also yielded organic substrates for SRBs at a moderate rate while MWTS and CHC at a much slower rate. The results show the unsuitability of BC as organic substrate for SRBs.

ORP: The initial ORP value in 5 batch tests was in the range of +85 to +130 mV (Figure 3.2c). ORP values in the batches with RC and PWTS as organic substrates reached below -200 mV in 2 d and continued to decrease during the experiment. ORP values with CHC and MWTS reached -100 mV in 6 and 13 d, respectively. In contrast, the batch with BC as organic substrate did not reach anaerobic conditions as evidenced by the positive redox values.

Sulfate reduction: The progress of sulfate reduction in single substrate batch tests followed a similar pattern as shown in Figure 3.2d. From Figure 3.2d, in the first 2 d of the batch test, the sulfate concentration increased as compared to the initial value due to leaching out of sulfate from the organic substrate. From day 4 onwards, the drop in the sulfate concentration was evident indicating that the SRB population was established and growing. This was also confirmed from the negative value of the ORP from day 2 onwards (Figure 3.2c), indicating the existence of anaerobic conditions necessary for growth of SRBs. The downward trend in sulfate concentration continued throughout the remaining batch period up to day 16.

Table 3.4. VS/TS ratio of the single substrate batch tests

Organic Substrate	VS/TS on Day 1	VS/TS on Day 16	Relative Rate of Biodegradation
BC	0.984	0.984	Slow
MWTS	0.455	0.412	Fast
PWTS	0.625	0.594	Fast
RH	0.788	0.763	Moderate
CHC	0.957	0.957	Slow

The sulfate reduction achieved at the end of the batch tests with 5 organic substrates is presented in Figure 3.2e. Comparing sulfate concentrations on day 2 and day 16, the batch tests with RH as the organic substrate recorded the highest sulfate removal efficiency (77.8%), followed by PWTS (66.7%), MWTS (60%), CHC (36.1%) and BC (30.3%).

Table 3.5. Comparative analysis of parameters on the last day (the 16th day) of the single substrate batch tests

Organic Substrate	pH	Alkalinity (mg L^{-1})	ORP (mV)	Sulfate reduction (%)	Suitability as Substrate for SRB
BC	4.39	0	+57	30.3±0.1	Not suitable
MWTS	6.84	612	-132	60.4±3.4	Suitable
PWTS	7.38	4150	-377	66.7±1.6	Suitable
RH	6.47	1327	-306	77.8±6.5	Suitable
CHC	6.93	159	-185	36.1±0.7	Suitable

Table 3.5 shows that all the organic substrates except BC were able to maintain the pH in the range 5-8 suitable for growth of SRB (Gibert et al., 2002); could generated alkalinity, which is an indicator of biological sulfate reduction (Brown et al., 2002); and could also maintain the ORP in a negative range indicating the existence of anaerobic conditions necessary for SRB (Gibert et al., 2002). All organic substrates except BC yielded a significant percentage of sulfate reduction although CHC recorded a lower percentage of sulfate reduction than MWTS, PWTS, and RH. Likewise, the VS/TS data presented in Table 3.4 also show that MWTS, PWTS, and RH recorded a decrease in VS/TS from the beginning of the batch test (day 1) to the last day of the batch test (day 16). This signifies the availability of organic carbon as electron donor for sulfate reduction by SRB. On the other hand, although CHC did not record a reduction in

VS/TS ratio, the data in Table 3.5 show its suitability for sulfate reduction. Therefore, it can be concluded that all organic substrates except BC are suitable as substrates for SRB. PWTS is a fast degrading organic substrate, RH is moderately degrading and CHC is a slow degrading organic substrate. Thus, PWTS, RH and CHC were selected for multiple substrates batch tests.

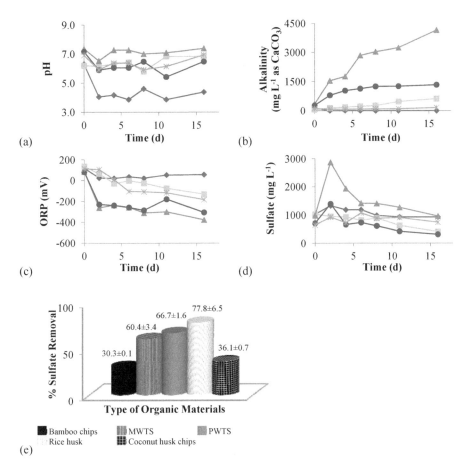

Figure 3.2. Individual substrate batch test (a) pH, (b) alkalinity, (c) ORP, (d) sulfate removal *vs* time and (e) sulfate removal efficiency (%) comparing sulfate concentrations on day 2 and day 16. (♦) BC, (■) MWTS, (▲) PWTS, (●) RH and (✳) CHC.

Researchers have reported the lignin content in various parts of a bamboo plant (Lybeer & Koch, 2005). According to the study of Chandler et al. (Chandler et al., 1980), the biodegradable fraction can be used to estimate the degradability of an organic substrate. A model equation to approximate the substrate biodegradable fraction (B) based on the lignin content was proposed as shown in the following equation:

$$B = -0.028X + 0.830 \tag{3.6}$$

where B is expressed on a VS basis and X is the lignin content of the VS, expressed as percent dry weight. Lignin, which is a complex phenolic polymer (Pouteau et al., 2003), serves an important function in plant defense due to its insolubility and complexity, which makes it resistant to degradation by most microorganisms (Campbell & Sederoff, 1996). In addition, Gibert et al. (Gibert et al., 2004) found that the lower the lignin content in the organic substrate, the higher its degradability and capacity for developing bacterial activity. Batch and column tests were conducted to correlate the lignin content of four different natural organic substrates (compost, sheep and poultry manures, and oak leaf) and their capacity to sustain bacterial activity. The organic substrates which resulted in high degradability and capacity for developing bacterial activity had low lignin content. Sheep manure had the lowest lignin content, followed by compost and oak leaf. Therefore, sheep manure was clearly the most suitable electron donor (sulfate removal level of > 99%) followed by poultry manure and oak leaf. The lignin content of the various substrates is presented in Table 3.2. Since PWTS and MWTS are from microbial and not plant origin materials, they have the lowest lignin content (He et al., 1998). On the other hand, RH, CHC and BC originate from plant materials, they have a significantly higher lignin content. Furthermore, single substrate batch tests showed that BH which had a high lignin content (Vu et al., 2003) were found to be the least suitable electron donor for growth of SRB.

3.3.2.2 Multiple substrates batch tests

Based on the results obtained from the single substrate batch test, multiple substrates batch tests were conducted for 40 d using 4 mixtures, namely: 1) RH + CHC; 2) PWTS + CHC; 3) PWTS + RH; and 4) PWTS + CHC + RH.

pH and alkalinity: The pH was maintained in the neutral range (pH 6-8) throughout the experiments (Figure 3.3a). There was a moderate increase in alkalinity in all four mixtures (Figure 3.3b).The highest alkalinity (4093 mg L^{-1} as $CaCO_3$) was produced in the PWTS + RH mixture, whereas the lowest alkalinity, (1287 mg L^{-1} as $CaCO_3$) was produced in the RH + CHC mixture.

ORP: The ORP of the AMD started from +115.3 mV and changed to different values after AMD was added into each organic material mixture (Figure 3.3c). The ORP of the liquid phase reached below -300 mV in 10 d and stabilized in the range between -300 and -376 mV at the end of the experiment.

Sulfate reduction: Throughout the single and multiple substrates batch tests, there was a significant fluctuation in the liquid phase sulfide concentration. Sulfide, which is a reaction product of the SRBs, is also one of the ionic products from the dissolution of H_2S gas as follows:

$$H_2S \leftrightarrow HS^- + H^+ \tag{3.7}$$

From this reaction, the concentration of sulfide is pH dependent (Sawyer et al., 2003). Under acidic conditions sulfide combines with protons to produce H_2S which may be released as gaseous product. Other researchers also have reported pH dependent loss of sulfur from volatilization of H_2S (Jong & Parry, 2003). As a result, it was more convenient to use the sulfate concentration data in these experiments to evaluate the sulfate reduction rate. The results were similar to the single material tests. The initial sulfate concentration in the liquid phase was higher than the sulfate in the AMD because of sulfate released from the media. After that, there was a decrease in the sulfate concentration as the experiment progressed. The RH + CHC mixture had the lowest sulfate reduction efficiency of 84% on day 32 (Figures 3.3d and 3.3f). The sulfate reduction efficiencies (comparing sulfate concentrations on day 2 and day 32) of the PWTS + RH, PWTS + RH + CHC and PWTS + CHC mixtures were slightly different (99%, 98%, and 95%, respectively).

From the first order rate equation, the sulfate reduction rate is proportional to the sulfate concentration. Rate constants (k) of RH + CHC, PWTS + CHC, PWTS + RH and PWTS + RH + CHC media were estimated at 0.121 d^{-1}, 0.196 d^{-1}, 0.277 d^{-1}, and 0.206 d^{-1}, respectively (Figure 3.3e).

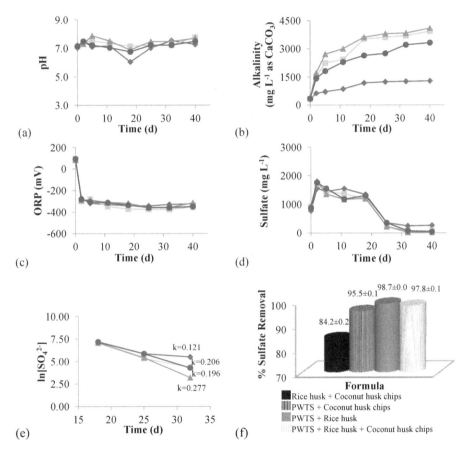

Figure 3.3. Mixed substrate batch test (a) pH, (b) alkalinity, (c) ORP, (d) sulfate removal *vs* time, (e) First-order rate constant for mixed substrates and (f) sulfate removal efficiency (%) comparing sulfate concentrations on day 2 and day 32.
(◆) RH + CHC, (■) PWTS + CHC, (▲) PWTS + RH and (●) PWTS + RH + CHC.

3.3.3 Continuous column experiments

3.3.3.1 Leaching tests and elemental analysis of organic substrates

Table 3.6 gives the elemental composition of the 3 organic substrates, namely: PWTS, RH and CHC prior to the start up of the column tests. PWTS, being excess biomass from a pig farm wastewater treatment plant, showed a higher proportion of N and metals (Fe, Cu, Zn and Mn) due to their accumulation in the biomass. RH and CHC, being of plant origin, showed a lower proportion of metals.

Table 3.6 shows that all the three organic substrates leached out metals, sulfate and DOC. PWTS consistently showed a higher concentration of heavy metals (Fe, Cu, Zn and Mn), sulfate and DOC in the leachate as compared to RH and CHC. PWTS tends to accumulate heavy metals and is more easily degradable; hence sulfate and DOC leach out in higher proportion. On the other hand, RH and CHC tend to reveal lower concentrations of metals, sulfate and DOC. These results also bring out the possibility that during the initial stages of the column experiments, metals and sulfate could be leached out from these organic substrates, particularly from PWTS. Likewise, PWTS generated a higher quantity of DOC which is essential for SRB as source of electron donor.

Table 3.6. Elemental composition and leachable amount from the organic substrates

Organic substrate		PWTS	RH	CHC
Elemental Composition (%)	C	27.31	30.65	46.02
	H	4.99	4.25	5.29
	N	3.57	1.15	1.01
	Fe	4.27	0.18	0.33
	Cu	0.13	0.01	0.02
	Zn	1.03	0.11	0.13
	Mn	0.80	0.11	0.01
Leachable amount (mg g^{-1})	Fe	0.467	0.057	0.065
	Cu	0.034	0.001	0.001
	Zn	0.253	0.036	0.012
	Mn	0.101	0.100	0.003
	SO_4^{2-}	9.227	1.347	1.273
	DOC	5.365	5.008	3.991

3.3.3.2 pH and alkalinity

Continuous column experiments exhibited the well-established pattern of an initial acclimatization period followed by growth of SRB leading to sulfate reduction. As the operation progressed, effluent pH increased, redox potential dropped and metal removal increased gradually.

SRBs require a pH in the range of 5-8 for their growth. Furthermore, sufficient alkalinity is also necessary in order to resist the acidity of AMD. Harris and Ragusa (Harris & Ragusa, 2000) observed that SRB can operate in waters at significantly lower pH values, when such waters were previously provided with effective pH buffers. A long adaptation period was needed for SRB to become active.

A deoxygenated AMD supplemented with heavy metals with characteristics as shown in Table 3.1 was continuously fed to the columns. At the end of the experiment, Reactor 2 with PWTS, RH and CHC in the ratio of 60:20:20 yielded the best result in terms of pH and alkalinity, with an average pH of 7.31 ± 0.04 (Figure 3.4a) and alkalinity of 503 ± 12 mg L^{-1} (Figure 3.4b). On the other hand, Reactor 3 which contained PWTS, RH and CHC in the ratio of 20:60:20 yielded a lower pH (6.82 ± 0.06) and a lower alkalinity (43 ± 5 mg L^{-1}).

3.3.3.3 ORP and DOC

Reactor 2 yielded the lowest ORP of -300 mV which is suitable for growth of SRB (Figure 3.4c). Reactor 3 yielded the worst performance with respect to ORP, it fluctuated around 0 mV, indicating no strict anaerobic conditions which may be unfavorable for growth of SRBs (Gibert et al., 2002).

The release of DOC during the degradation of organic material as source for energy and carbon for SRB is also necessary (Drury, 1999; Hammack & Edenborn, 1992). Supply of DOC from organic materials can be divided in two different types of reactions. First is the short-term elution of soluble organic molecules. In this process, microbial processes are regarded insignificant for DOC supply. Another type is active elution, which is the long-term release of DOC after hydrolysis and fermentation of macromolecular compounds by microbial processes. As far as the DOC in the effluent is concerned, Reactor 2 containing PWTS, RH and CHC in the ratio of 60:20:20 yielded a higher DOC value starting from about 900 mg L^{-1} at the beginning to about 35 mg L^{-1} at the end of the experiment (Figure 3.4d).

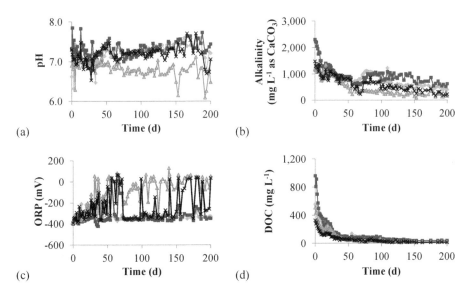

Figure 3.4. Effluent (a) pH, (b) alkalinity, (c) ORP and (d) DOC of column test.
(♦) Reactor 1, (■) Reactor 2, (▲) Reactor 3 and (×) Reactor 4.

3.3.3.4 Heavy metal removal

Heavy metal removal from the column experiments is presented in Figure 3.5. The results reveal that the four reactors achieved over 80% removal of Fe, Cu and Zn within the first 10 d of operation. These heavy metal removal efficiencies remained consistently high with Zn removal, which remained over 80% throughout the period of 200 d. However, the Mn removal efficiency was satisfactory (over 90%) only in reactor 2 while the other reactors showed a lower Mn removal efficiency.

Reactor 2 yielded the best Fe, Cu, Zn and Mn removal efficiency. More than 90% of all these heavy metals were removed in Reactor 2. Although Fe, Cu and Zn removal has been satisfactory in Reactor 1, Reactor 3 and Reactor 4, Mn removal in these reactors was not satisfactory. Out of these 3 reactors, Reactor 3 gave the lowest Mn removal efficiency.

The metal removal can be attributed to the precipitation of insoluble metal sulfides as a result of sulfide production from SRB activity in the continuous reactors. Cu removal was the most stable and efficient (Figure 3.5b), followed by Zn and Fe (Figures 3.5c and 3.5a, respectively). Metal removal from AMD in an experimental constructed wetland was found to follow closely the trend in solubility product (K_{sp}) values (Machemer & Wildeman, 1992). Log K_{sp} values of CuS, ZnS, FeS and MnS are -35.06, -20.96, -18.10 and -14.29, respectively (Chang, 2009). The metal removal trend in the continuous column experiment indeed followed this trend as Cu removal was the highest (Figure 3.5b), while the Mn (Figure 3.5d) removal was the lowest. Mn removal was least in Reactors 1, 3 and 4; Reactors 2 and 3 yielded the best and the worst Mn removal, respectively (Figure 3.5d). As the log K_{sp} of MnS was the highest of all the metal sulfides investigated in this research, not all Mn in the reactor precipitate as MnS and the remaining Mn was discharged in the effluent in its dissolved form.

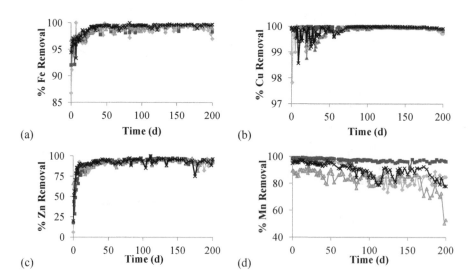

(a) (b) (c) (d)

Figure 3.5. Heavy metal removal efficiency (%) of column test.
($\blacklozenge$) Reactor 1, ($\blacksquare$) Reactor 2, ($\blacktriangle$) Reactor 3 and (×) Reactor 4.

The metal removal process can be both abiotic and biotic. Therefore, the metal removal could be a combination of sulfide precipitation and adsorption onto the organic substrates. The metal adsorption on the organic substrates was also tested. The results showed that up to 20% of the metals were adsorbed for RH and CHC (data not shown). However, to differentiate between metal removal by adsorption and sulfide precipitation was particularly difficult for the PWTS as it contained an active SRB population from the very beginning.

Reactor 2 which had PWTS, RH and CHC in the ratio of 60:20:20 had a higher proportion of PWTS which is a fast degrading organic substrate. Due to this, the DOC was readily available for SRBs as electron donor. Hence, Reactor 2 recorded a higher and more consistent heavy metal removal efficiency due to higher growth and activity of the SRB. Researchers have used single as well as multiple substrates as possible electron donor in PRBs (Costa et al., 2007; Gibert et al., 2008; Soares & Abeliovich, 1998). Since recharging of electron donor in a PRB under sub-surface conditions cannot be done frequently, use of multiple substrates offers an attractive alternative (Waybrant et al., 1998). Such mixtures of multiple substrates with slow, moderate and fast degrading electron donors would make DOC available as electron donor for SRBs in early stages from fast degrading organic substrates and in longer duration from slow degrading organic substrates. Reactor 2, which employed a mixture of PWTS, RH and CHC in the ratio of 60:20:20 yielded the best heavy metal removal efficiency. On the other hand, the heavy metal removal performance of Reactor 1 (RH and CHC in the ratio of 33:33:33) was also satisfactory. As compared to this, Reactor 3 (PWTS, RH and CHC in the ratio of 20:60:20) and Reactor 4 (PWTS, RH and CHC in the ratio of 20:20:60) had a lower proportion of fast degrading organic substrates. These reactors consistently yielded a lower heavy metal removal efficiency.

3.4. Conclusions

This investigation demonstrated biological sulfate reduction and subsequent sulfide precipitation of Cu, Zn, Fe and Mn by mixed populations of SRB in batch as well as continuous columns containing a variety of organic substrates. The following conclusions can be drawn from this investigation:

- Batch experiments showed that both mixtures of PWTS + RH and PWTS + RH + CHC yielded better conditions for sulfate reduction. Both mixtures developed conditions (generation of alkalinity and a low ORP) that favor the activity and growth of SRB leading to biological sulfate reduction.
- Continuous column experiments showed that a mixture of PWTS, RH and CHC could successfully facilitate growth of SRB and yielded above 95% removal of Cu, Zn and Fe. As far as Mn removal was concerned, a reactor column with PWTS, RH and CHC in a proportion of 60:20:20 yielded a Mn removal efficiency exceeding 95%, while in other reactor columns, Mn removal was not satisfactory.
- It is recommended that a mixture of fast and slow degrading organic substrates such as PWTS, RH and CHC are utilized in PRBs as electron donor for growth of SRB for removal of heavy metals from AMD.

3.5 References

Bilba, K., Arsene, M., & Ouensanga, A. (2007). Study of banana and coconut fibers: Botanical composition, thermal degradation and textural observations. *Bioresour. Technol., 98*, 58-68.

Blasi, C. D., Buonanno, F., & Branca, C. (1999). Reactivities of some biomass chars in air. *Carbon, 37*, 1227-1238.

Brown, M., Barley, B., & Wood, H. (2002). *Minewater Treatment*. London: IWA Publishing.

Campbell, M. M., & Sederoff, R. R. (1996). Variation in lignin content and composition: Mechanisms of control and implications for the genetic improvement of plants. *Plant Physiol., 110*, 3-13.

Chandler, J. A., Jewell, W. J., Gossett, J. M., van Soest, P. J., & Robertson, J. B. (1980). Predicting methane fermentation biodegradability. *Biotechnol. Bioeng. Symp., 10*, 93-107.

Chang, I. S., Shin, P. K., & Kim, B. H. (2000). Biological treatment of acid mine drainage under sulphate-reducing conditions with solid waste materials as substrate. *Water Res., 34*(4), 1269-1277.

Chang, R. (2009). *Chemistry* (10th ed.). U.S.A.: McGraw Hill.

Christensen, B., Laake, M., & Lien, T. (1996). Treatment of acid mine water by sulfate reducing bacteria: Results from bench scale experiment. *Water Res., 30*(7), 1617-1624.

Costa, M. C., Martins, M., Jesus, C., & Duarte, J. C. (2007). Treatment of acid mine drainage by sulfate-reducing bacteria using low cost matrices. *Water Air Soil Pollut., 189*(1-4), 149-162.

Drury, W. J. (1999). Treatment of acid mine drainage with anaerobic solid substrate reactors. *Water Environ. Res., 71*, 1244-1250.

Dvorak, D. H., Hedin, R. S., Edenborn, H. M., & McIntire, P. E. (1992). Treatment of metal-contaminated water using bacterial sulfate reduction: Results from pilot-scale reactors. *Biotechnol. Bioeng., 40*(5), 609-616.

Eaton, A. D., APHA, AWWA, & WEF. (2005). *Standard methods for the examination of water and wastewater* (21st ed.). Washington D.C.

Elliott, P., Ragusa, S., & Catcheside, D. (1998). Growth of sulfate-reducing bacteria under acidic conditions in an upflow anaerobic bioreactor as a treatment system for acid mine drainage. *Water Res., 32*(12), 3724-3730.

Gibert, O., de Pablo, J., Cortina, J. L., & Ayora, C. (2002). Treatment of acid mine drainage by sulfate-reducing bacteria using permeable reactive barrier: A review from laboratory to full-scale experiments. *Rev. Environ. Sci. Biotechnol., 1*(4), 327-333.

Gibert, O., de Pablo, J., Cortina, J. L., & Ayora, C. (2004). Chemical characterization of natural organic substrates for biological mitigation of acid mine drainage. *Water Res., 38*, 4186-4196.

Gibert, O., Pomierny, S., Rowe, I., & Kalin, R. M. (2008). Selection of organic substrates as potential reactive materials for use in a denitrification permeable reactive barrier (PRB). *Bioresour. Technol., 99*(16), 7587-7596.

Gibert, O., Rotting, T., Cortina, J. L., de Pablo, J., Ayora, C., & Carrera, J. (2011). In-situ remediation of acid mine drainage using a permeable reactive barrier in Aznalcollar (Sw Spain). *J. Hazard. Mater., 191*, 287-295.

Gray, N. F. (1997). Environmental impact and remediation of acid mine drainage: a management problem. *Environ. Geol., 30*(1-2), 62-71.

Hammack, R. W., & Edenborn, H. M. (1992). The removal of nickel from mine waters using bacterial sulfate reduction. *Appl. Microbiol. Biotechnol., 37*, 674-678.

Harris, M. A., & Ragusa, S. (2000). Bacterial mitigation of pollutants in acid drainage using decomposable plant material and sludge. *Environ. Geol., 40*, 195-215.

He, B. J., Zhang, Y., Riskowski, G. L., & Funk, T. L. (1998). *Thermochemical conversion of swine manure: Temperature and pressure responses.* Paper presented at the ASAE Annual International Meeting.

Huttagosol, P., & Kijjanapanich, V. (2008). *Active Chemical Treatment of Water in the Open Pit of Ban Hong Coal Mine*: Metallurigical and Petroleum Technology Development, Chulalongkorn University.

Jong, T., & Parry, D. L. (2003). Removal of sulfate and heavy metals by sulfate-reducing bacteria in short term bench scale upflow anaerobic packed bed reactor runs. *Water Res., 37*, 3379-3389.

Keely, J. F. (1989). Performance evaluations of pump-and-treat remediations. *U.S. EPA. Superfund Ground Water Issue, EPA/540/4-89/005.*

Lapointe, F., Kostas, F., & McConchie, D. (2006). Efficiency of Bauxsol^{TM} in permeable reactive barriers to treat acid rock drainage. *Mine Water Environ., 25*, 37–44.

Levenspiel, O. (1999). *Chemical reaction engineering* (3rd ed.). U.S.A.: John Wiley & Sons, Inc.

Lybeer, B., & Koch, G. (2005). Lignin distribution in the tropical bamboo species gigantochloa levis. *IAWA J., 26*(4), 443–456.

Machemer, S. D., & Wildeman, T. R. (1992). Adsorption compared with sulfide precipitation as metal removal processes from acid mine drainage in a constructed wetland. *J. Contam. Hydrol., 9*(1-2), 115-131.

Masscheleyn, P. H., Delaune, R. D., & Patrick, J. W. H. (1991). Effect of redox potential and pH on arsenic speciation and solubility in a contaminated soil. *Environ. Sci. Technol., 25*, 1414-1419.

Morrison, S. J., & Spangler, R. R. (1992). Extraction of uranium and molybdenum from aqueous solutions: A survey of industrial materials for use in chemical barriers for uranium mill tailings remediation. *Environ. Sci. Technol., 26*(10), 1922-1931.

Morrison, S. J., & Spangler, R. R. (1993). Chemical barriers for controlling groundwater contamination. *Environ. Prog., 12*, 175-181.

National-Research-Council. (1994). *Alternatives for Ground Water Cleanup.* Washington, D. C: National Academy Press.

Ngwenya, E. G., Maud, W., Caroline, A. M.-D., & Ralph, J. P. (2006). Remediation of acid mine drainage leachate by a physicochemical and biological treatment-train approach. *Remediation Journal, 6*(2), 79-90.

Pagnaneli, F., Viggi, C. C., Mainlli, S., & Toro, L. (2009). Assessment of solid reactive mixtures for the development of biological permeable reactive barriers. *J. Hazard. Mater., 170*, 998-1005.

Pouteau, C., Dole, P., Cathala, B., Averous, L., & Boquillon, N. (2003). Antioxidant properties of lignin in polypropylene. *Polym. Degrad. Stab., 81*, 9-18.

Sawyer, C. N., McCarty, P. L., & Parkin, G. F. (2003). *Chemistry for environmental engineering and science* (5th ed.): Mc Graw-Hill International

Soares, M. I. M., & Abeliovich, A. (1998). Wheat straw as substrate for water denitrification. *Water Res., 32*, 3790-3794.

Stumm, W., & Morgan, J. J. (1981). *Aquatic chemistry: An introduction emphasizing chemical equilibria in natural waters* (2nd ed.): John Wiley & Sons, Inc.

Tsukamoto, T. K., Killion, H. A., & Miller, G. C. (2004). Column experiments for microbiological treatment of acid mine drainage: low-temperature, low pH and matrix investigations. *Water Res., 38*, 1405-1418.

Vu, T. H. M., Pakkanen, H., & Alén, R. (2003). Delignification of bamboo (Bambusa procera acher) Part 1. Kraft pulping and the subsequent oxygen delignification to pulp with a low kappa number. *Ind. Crops Products, 19*, 49-57.

Walton-Day, K. (2003). *Passive and active treatment of mine drainage: Environmental Aspects of Mine Wastes*. Ottawa: Mineralogical Assoc of Canada.

Waybrant, K. R., Blowes, D. W., & Ptacek, C. J. (1998). Selection of Reactive Mixtures for Use in Permeable Reactive Walls for Treatment of Mine Drainage. *Environ. Sci. Technol., 32*(13), 1972-1979.

Zheljazkov, D. V., & Nielsen, N. E. (1996). Effect of heavy metals on peppermint and cornmint. *Plant and Soil, 178*, 59-66.

CHAPTER 4

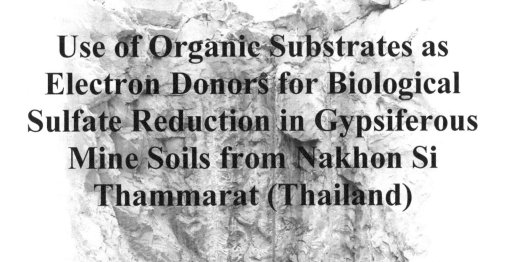

Use of Organic Substrates as Electron Donors for Biological Sulfate Reduction in Gypsiferous Mine Soils from Nakhon Si Thammarat (Thailand)

This chapter has been published as:
Kijjanapanich, P., Annachhatre, A. P., Esposito, G., & Lens, P. N. L. (2013). Use of organic substrates as electron donors for biological sulfate reduction in gypsiferous mine soils from Nakhon Si Thammarat (Thailand). *Chemosphere, In Press.*
DOI: 10.1016/j.chemosphere.2013.11.026

Chapter 4

Soils in some mining areas contain a high gypsum content, which can give adverse effects to the environment and may cause many cultivation problems, such as a low water retention capacity and low fertility. The quality of such mine soils can be improved by reducing the soil's gypsum content. This study aims to develop an appropriate *in situ* bioremediation technology for abbreviating the gypsum content of mine soils by using sulfate reducing bacteria (SRB). The technology was applied to a mine soil from a gypsum mine in the southern part of Thailand which contains a high sulfate content (150 g kg^{-1}). Cheap organic substrates with low or no cost, such as rice husk, pig farm wastewater treatment sludge and coconut husk chips were supplied to the soil as electron donors for the SRB. The highest sulfate removal efficiency of 59% was achieved in the soil mixed with 40% organic mixture, corresponding to a reduction of the soil gypsum content from 25% (pure gypsum mine soil) to 7.5%. For economic gains, this treated soil can be further used for agriculture and the produced sulfide can be recovered as the fertilizer elemental sulfur.

4.1. Introduction

Soils containing significant quantities of gypsum, which may interfere with plant growth, are defined as gypsiferous soils (FAO, 1990). In the natural environment, gypsum can be transported by water or wind, be re-deposited at new locations forming individual gypsum dunes or it can be incorporated in the soil layer (FAO, 1990). The main reason for gypsum accumulation in the soil is its precipitation from supersaturated underground or runoff waters, as a result of intensive evaporation. Gypsum is also formed in acid sulfate soils (Dent, 1986). In these soils, the origin of the sulfate ions (SO_4^{2-}) is due to the oxidation of sulfur rich minerals such as pyrite (FeS_2) in the parent material. Due to natural weathering and oxidation cycles, the sulfur in these minerals is transformed into sulfuric acid, causing calcareous soils to react with calcium carbonate ($CaCO_3$) forming gypsum (Dent, 1986; FAO, 1990).

Gypsiferous soils have received little curative attention as compared to most other affected soil types, and have been considered to have little or no agricultural potential (FAO, 1974; USDA, 1975). The presence of gypsum in gypsiferous soils creates several problems for their agricultural use and development, including low water retention capacity, shallow depth to the hardpan and vertical crusting (Khresat et al., 2004). The accumulation of gypsum in soils results in very low fertility, and consequently, their productivity remains low under irrigation even with application of fertilizers or organic manures (FAO, 1990). With this kind of soils, larger amounts of phosphorous application are needed because of the greater phosphorus immobilization by the gypsum (Verheye & Boyadgiev, 1997). Compared to a non-gypsiferous soil, the amount of the calcium and sulfate ions in the soil solution is increased due to the solubility of gypsum, resulting in calcite precipitation (Kordlaghari & Rowell, 2006). The impact of these adverse properties depends on the gypsum content and the depth at which the gypsiferous layer occurs in the root zone (Verheye & Boyadgiev, 1997). Under saturated conditions, gypsum may impregnate most of the soil matrix. When less calcium sulfate is present in the system, gypsum precipitates in localized spots (Verheye & Boyadgiev, 1997).

The physical structure of gypsiferous soils such as its porosity and permeability can be improved by reducing the soil's gypsum content (Alfaya et al., 2009). A gypsum content of 2-10% does not interfere significantly with the soil structure. The gypsum crystals, however, tend to break the continuity of the soil mass in soils which contain 10-25% of gypsum. Soils with more than 25% gypsum are considered unsuitable for most crops. Under such conditions, gypsum may precipitate and can cement soil material into hard layers, thus roots cannot penetrate except for those of very tolerant crops such as alfalfa, clover or oats (Smith & Robertson, 1962; Verheye & Boyadgiev, 1997).

The problems mentioned above also occur in several mining areas, especially gypsum mines, where the soils have a high gypsum content and cannot be used for agriculture. For instance, soils in the gypsum mine in the southern part of Thailand (Figure 4.1a) have a high sulfate content that can induce adverse effects on the environment. Moreover, the soils of some mines can also generate acid mine drainage (AMD) and mass mortalities of plants and aquatic life (Kijjanapanich et al., 2012). This AMD has a low pH and high concentrations of sulfate and toxic metals. Such land cannot be used for agriculture, and these soils have a poor fauna and flora.

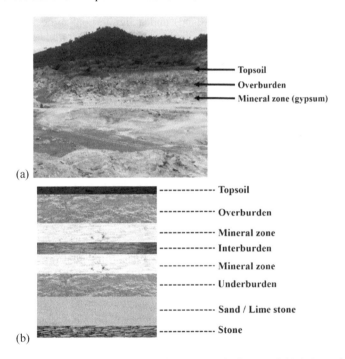

Figure 4.1. Mining: (a) Gypsum mine in Nakhon Si Thammarat, Thailand and (b) Schematic representation of soil profile in a mining zone.

Little research has been done on the bioremediation of gypsiferous soils. Alfaya et al. (2009) ascertained that calcareous gypsiferous soils contain an endogenous SRB population that uses the sulfate from gypsum in the soil as an electron acceptor. The sulfate reduction rate doubled when anaerobic granular sludge was added to bioaugment the soil with SRB. In the presence of anaerobic granular sludge, a maximum sulfate reduction rate of 567 mg L^{-1} d^{-1} was achieved with propionate as the electron donor.

Most of the gypsiferous soils have a relatively low organic matter content (Ghabour et al., 2008). Therefore, appropriate electron donor needs to be added for the SRB when designing a bioremediation scheme for gypsiferous soils based on biological sulfate reduction.

This research aimed to study the characteristics of soils from a lignite coal mine and a gypsum mine. Gypsiferous soils from a gypsum mine (Figure 4.1a), containing a high gypsum content, was treated by biological sulfate reduction (batch experiments) in order to reduce the gypsum content by using no or low cost organic substrates as electron donors for SRB.

4.2. Material and Methods
4.2.1 Mine soils (overburdens)

Two different types of soil samples were used in this study: gypsum mine overburden (GMOB) and lignite coal mine overburden (LMOB). The overburdens of a mine are the rock and soil part that lies above the ore body and needs to be excavated by open pit mining (Figure 4.1b). GMOB and LMOB were collected from a gypsum mine in Nakhon Si Thammarat (Thailand) and a lignite coal mine in Lam Phun (Thailand), respectively. All samples were air-dried and sieved at 2 mm. These overburden samples were then analyzed for pH, soil texture, organic matter (OM), cation-exchange capacity (CEC), synthetic precipitation leaching procedure (SPLP) and waste extraction test (WET).

4.2.2 Sulfate reducing bacteria (SRB) inoculums

Sludge from a pilot scale mesophilic anaerobic channel digester and upflow anaerobic sludge blanket (UASB) reactor treating pig farm wastewater operated at the Energy Research and Development Institute-Nakonping, Chiang Mai University (Thailand) was used as source of SRB. The seed sludge had a TSS and VSS content of 33.3 g L^{-1} and 21.3 g L^{-1}, respectively, corresponding to a VSS/TSS ratio of 0.64.

4.2.3 Organic substrates

Three types of organic substrates were selected for their possible use as electron donor for SRB (Kijjanapanich et al., 2012). These included rice husk (RH), pig farm wastewater treatment sludge (PWTS) and coconut husk chips (CHC), based on their availability, ease of handling and no or low cost. Organic substrates were air dried, cut to the desired size and analyzed for their physical characteristics. The physical characteristics of these organic substrates was described in the study of Kijjanapanich et al. (2012). PWTS, RH and CHC were mixed in a ratio of 60:20:20 (by volume) (Kijjanapanich et al., 2012) prior to use.

4.2.4 Column leaching experiments

The leaching columns had a working volume of 2 L and were made of polyvinyl chloride (PVC) and filled with one kg of GMOB or LMOB in each leaching column. They were operated at room temperature (25 ± 5°C) for 28 and 32 d for LMOB and GMOB, respectively. Demineralized water was fed at the bottom of the column at a flow rate of 252 mL hr^{-1} (0.1 m hr^{-1}) using a peristaltic pump. The leachate was withdrawn from the top of the column and collected daily for analysis.

4.2.5 Bioreactor experiments

GMOB with a high sulfate content (around 150 g kg^{-1}), classified as gypsiferous soils, was selected for the bioreactor experiment. The reactors (working volume of 5 L) were made of polyethylene (PE) and operated at room temperature ($25 \pm 5°C$) (Figures 4.2a and 4.2b). Each reactor had a biogas releasing tube at the top. This hydrogen sulfide (H$_2$S) rich biogas was lead through a zinc acetate solution (1 M) in order to trap H$_2$S. Nitrogen gas was used to purge the reactor prior to sampling in order to make sure that most of the H$_2$S was trapped in the zinc acetate solution.

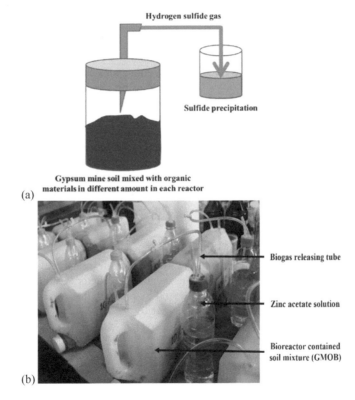

Figure 4.2. Schematic representation of the bioreactor used in this experiment: (a) reactor schematic and (b) lab-scale bioreactor.

GMOB (2500 g) was mixed with the organic mixture in different amounts: 10, 20, 30 and 40% of the GMOB, respectively. A SRB inoculum of 250 g (10% of the GMOB) was added to each bioreactor. Table 4.1 shows the composition of each soil mixture. During the acclimatization period, biological sulfate reduction may progress at a considerably slow pace, leading to a lower alkalinity generation. It was therefore necessary to adjust the pH of the soils to the optimum range for SRB (pH 6-7) at the beginning of the experiment. This was done by adding lime (0.02% of the GMOB) to the GMOB. Water was added to the reactors in order to maintain the desired moisture content (20-25%). Samples were collected once a week for analyzes.

Table 4.1. Composition of each soil mixture applied in this study

Mixture (%)	GMOB (g)	PWTS (g)	RH (g)	CHC (g)	Total mixture (g)	Sludge (g)
0	2500	-	-	-	2500	250
10	2500	196.9	44.7	8.4	2750	250
20	2500	393.8	89.3	16.9	3000	250
30	2500	590.7	134.0	25.3	3250	250
40	2500	787.6	178.7	33.7	3500	250

GMOB: gypsum mine overburden, PWTS: pig farm wastewater treatment sludge, RH: rice husk, CHC: coconut husk chips

4.2.6. Analytical methods

The pH was measured as overall acidity indicator using a Mettler Toledo pH meter. The Electro-Conductivity (EC) was measured using a HANNA HI 9835 conductivity meter. The leaching potential of GMOB and LMOB was measured using the Synthetic Precipitation Leaching Procedure (SPLP) (U.S.EPA, 1994). This method is applicable for materials where the leaching potential due to normal rainfall is to be determined. Instead of the landfill leachate simulating acetic acid mixture, sulfuric and nitric acids (60:40 weight percent mixture) pH 4.20 ± 0.05 are utilized in this study in an effort to simulate the acid rains resulting from airborne nitric and sulfuric oxides. The mobility of specific inorganic and organic contaminants that are destined for disposal in municipal landfills was estimated using the Waste Extraction Test (WET) (CA WET, 1984).

Sulfate was measured by the gravimetric method (sulfate concentrations above 10 mg L^{-1}) and the turbidimetric method (sulfate concentrations in the range of 1–40 mg L^{-1}) using a CECIL CE2030 UV visible spectrophotometer (Eaton et al., 2005). Sulfide was measured using the gravimetric method by precipitation as zinc sulfide in a zinc acetate solution (1 M). Calcium was measured by the EDTA titration method. Heavy metals (Mn, Zn, Cu and Fe) were measured using an Ananta MUA/USEEP Atomic Absorption Spectrometer (AAS)-Flame (Eaton et al., 2005). Fluoride and aluminium were measured by the SPADNS (sodium 2-(p-sulfophenylazo)-1,8-dihydroxynaphthalene-3,6-disulfonate) method (Eaton et al., 2005) and the Eriochrome cyanine R method (Eaton et al., 2005), respectively.

Soil texture was analyzed using the hydrometer method (Pansu & Gautheyrou, 2006). CEC was measured using the ammonium acetate method (Pansu & Gautheyrou, 2006). OM was measured using the Walkley & Black method (Walkley & Black, 1934). Total nitrogen, phosphorous and potassium were measured using the Kjeldahl method (Pansu & Gautheyrou, 2006), Bray II Extraction (Bray & Kurtz, 1945) and Atomic Emission Spectroscopy (AES) (Eaton et al., 2005), respectively.

4.3. Results

4.3.1 Mine soils (overburdens) characteristics

Table 4.2 shows the characteristics of GMOB and LMOB, which can both be classified as silt loam soils. The pH of the GMOB and LMOB was 4.9 and 3.3, respectively. The GMOB has a low organic matter content (1%), while the LMOB has a high organic matter content (17%).

Table 4.2. Characteristics of gypsum mine and lignite coal mine soils (overburdens)

	Parameters	GMOB	LMOB
Texture:	Sand, %	43.46	5.16
	Silt, %	50.78	69.08
	Clay, %	5.76	25.76
	Type	Slit loam	Slit loam
pH		4.89	3.31
Organic matter (OM), %		1.01	17.19
CEC, meq/100g		7.94	13.14

GMOB: gypsum mine overburden, LMOB: lignite coal mine overburden

The results of the SPLP and WET are shown in Table 4.3. The sulfate content in the GMOB was more than 15 times higher than in the LMOB. The GMOB has a high sulfate, calcium and iron content (146, 32 and 0.33 mg g^{-1}, respectively) and was selected for treatment in the bioreactor experiments.

Table 4.3. Synthetic precipitation leaching procedure (SPLP) and waste extraction test (WET) of gypsum mine and lignite coal mine soils (overburdens)

Parameters	GMOB		LMOB		Surface water quality standard of Thailand (1994)	STLC regulatory limits (CA WET, 1984)
	SPLP	WET	SPLP	WET		
pH	6.63	6.90	3.73	3.78	5.5-9.0	-
Sulfate (mg L^{-1})	1661.2	14633.4	379.0	964.1	-	-
Calcium (mg L^{-1})	1529.0	3270.8	90.1	200.2	-	-
Manganese (mg L^{-1})	0.322	2.978	2.770	10.963	<1.0	-
Zinc (mg L^{-1})	n.d.	0.563	0.668	2.407	<1.0	250
Copper (mg L^{-1})	n.d.	0.127	n.d.	0.437	<0.1	25
Iron (mg L^{-1})	2.172	32.767	0.139	10.608	-	-
Fluoride (mg L^{-1})	n.d.	48.59	n.d.	63.30	-	180
Aluminium (mg L^{-1})	2.29	n.d.	2.21	n.d.	-	-

GMOB: gypsum mine overburden, LMOB: lignite coal mine overburden, n.d.: not detected, STLC: Soluble Threshold Limit Concentration

4.3.2 Column leaching experiments

There was a slow increase in both leachate pHs. The pH of the GMOB and LMOB leachate increased from 5.9 to 7.9 and 3.7 to 5.0, respectively (Figure 4.3a). The EC of the LMOB leachate gradually decreased from 556 to 13 µS cm^{-1}, while the EC of the GMOB leachate fluctuated between 1700-2400 µS cm^{-1} during the first 16 d and then rapidly decreased from day 17 onwards to 60 µS cm^{-1} (Figure 4.3b).

The sulfate concentration of the leachate from the GMOB was in the range between 3-4129 mg L^{-1}, while the leachate from the LMOB contained 3-690 mg L^{-1} sulfate (Figure 4.3c). Up to 180 g and 10 g of sulfate were removed from 1 kg of GMOB and LMOB during 32 and 28 d of leaching, respectively. The dissolution of calcium followed the same pattern as the dissolution of sulfate (Figure 4.3d). Around 32 g and 2 g of calcium were removed from 1 kg of GMOB and LMOB during 32 and 28 d of leaching, respectively.

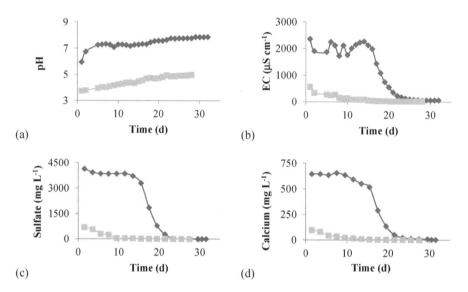

Figure 4.3. Evolution of the chemical composition of the leachate over leaching time: (a) pH, (b) EC, (c) sulfate and (d) calcium. (♦) Gypsum mine overburden and (■) Lignite coal mine overburden.

4.3.3 Bioreactor experiments

The GMOB had a very low OM (1%) and total nitrogen (0.05%) content, and contained 25% of gypsum prior to the treatment. The OM, total nitrogen, phosphorus and potassium content of the GMOB increased considerable after mixing with the organic mixture and remained at a high level after treatment (Table 4.4). Mixing of GMOB and the organic mixture created some dilution (up to 28% in case of 40% organic mixture) of the gypsum content of GMOB (Figure 4.4). However, the biological sulfate reduction process is still the main process of sulfate removal from the system and the dilution value is less when compared to the reduction of the gypsum content of GMOB by SRB (Figure 4.4).

The generated sulfide started to increase at week 3 and the maximum amount of sulfide was achieved between week 3 and 4 (Figure 4.5a). After week 4, the amount of sulfide started to reduce. The reactor with a 40% organic mixture yielded the highest amount of sulfide (14 g wk^{-1}). Sulfate in the soils also started to decrease at week 3 (Figure 4.5b). The lowest sulfate concentration (42 g kg^{-1}) was achieved in the 40% organic mixture reactor (Figure 4.5b and Table 4.4), corresponding to a sulfate removal efficiency of 59%. The calcium content of all soil mixtures remained constant throughout the experiment (Table 4.4).

Table 4.4. The characteristics of gypsum mine overburden (GMOB) before and after 105 d treatment

Percentage of organic substrate	0%		10%		20%		30%		40%	
Parameters	1	2	1	2	1	2	1	2	1	2
Sulfate (SO_4^{2-}), g kg^{-1}	131	138	121	84	111	80	104	56	102	42
Calcium (Ca), g kg^{-1}	33.8	34.0	32.3	30.2	30.6	30.4	29.5	28.7	25.0	26.2
OM, %	1.01	1.04	2.57	1.72	3.46	1.99	4.58	3.08	6.12	3.66
Total nitrogen, %	0.05	0.03	0.13	0.13	0.21	0.20	0.28	0.27	0.34	0.29
Available P, mg kg^{-1}	32.8	33.0	233.0	209.2	467.8	379.6	746.2	706.0	881.2	662.5
Exchangeable K, mg kg^{-1}	155.3	154.3	171.7	171.8	188.2	185.8	204.6	202.8	211.0	194.7

OM: Organic matter, P: Phosphorus, K: Potassium, 1: Before, 2: After 105 d

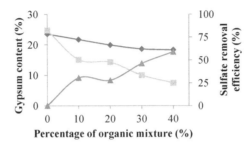

Figure 4.4. Performance of the gypsum removal from GMOB by biological sulfate reduction using different percentage of organic mixture for 105 d treatment. (♦) GMOB before treated, (■) GMOB after treated and (▲) gypsum removal efficiency (by sulfate reduction).

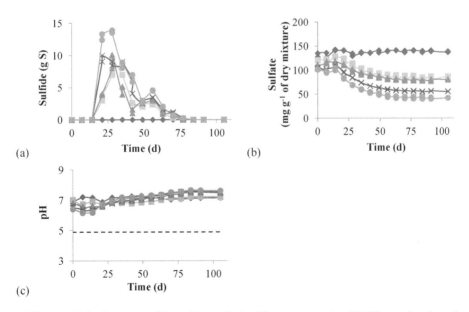

Figure 4.5. Performance of the sulfate reducing bioreactor treating GMOB as a function of operation time: (a) sulfide, (b) sulfate and (c) pH. (-) original soil and addition of (♦) 0%, (■) 10%, (▲) 20%, (×) 30% and (●) 40% of organic mixture.

The pH of the soils which contained organic substrates deceased during the starting phase of the experiments (Figure 4.5c). However, when sulfate reduction commenced from week 3 onwards, the pH of the soils with organic material continued to increase. The soil mixed with the 40% organic mixture had the highest pH of 7.6 at the end of the experiment (Figure 4.5c). In contrast, the pH of the soils without organic material was almost constant at pH 7 due to the addition of lime at the beginning of the experiment (Figure 4.5c).

Table 4.4 shows the characteristics of the GMOB before and after treatment. A change in color of all GMOB mixtures was observed. The color of GMOB mixtures became darker with time during the treatment with the organic material supplementation. The intensity of the color of the soil was related with the percentage of the organic material added to the soil. The soil supplemented with 40% organic mixture had the darkest color and became black already within one week of incubation.

4.4. Discussion
4.4.1 Characteristics of the leachate of mine soils (overburdens)

The GMOB had a very low OM (1%) and a high gypsum (25%) content, which can be classified as gypsiferous soils (FAO, 1990). This study showed that GMOB has a higher sulfate, calcium and iron content as compared to LMOB. According to the WET results (Table 4.3), heavy metals contained in both leachates did not exceed the soluble threshold limit concentration (STLC) (CA WET, 1984). Thus, GMOB and LMOB can be defined as non-hazardous material. However, LMOB can be sources of AMD generation, with a leachate that has a low pH (pH 3.3) and high manganese concentration (2.8 mg L^{-1}) (Pollution Control Department of Thailand, 1994). Therefore, technologies for remediation of the AMD generation from these mine soils, such as surface packing (Johnson & Hallberg, 2005), electrokinetic remediation (Acar et al., 1995; Virkutyte et al., 2002) or soils washing (Moutsatsou et al., 2006), should be considered and studied.

Gypsum has a solubility of 2600 mg L^{-1} in pure water at 25 °C (FAO, 1990), resulting in a sulfate concentration of 1450 mg L^{-1}. The highest sulfate concentration found in this study was 4129 mg L^{-1} in the GMOB leachate (Figure 4.3c). The sulfate concentration of the leachate exceeding the solubility limit can indicate that the leachate samples were supersaturated with sulfate. Jang and Townsend (2001) found that the sulfate concentrations exceeded the solubility limit from gypsum dissolution (up to 1585 mg L^{-1}) in many of the construction and demolition debris (CDD) leachate samples. Also phosphogypsum leachate can contain sulfate concentrations exceeding the gypsum solubility (Battistoni et al., 2006). This is possible due to the presence of other ions and the increased ionic strength of the leachate (Jang & Townsend, 2001). For instance, the gypsum solubility was found to be 3 times higher in the presence of 5 g L^{-1} sodium (Shternina, 1960). Supersaturation of sulfate can also occur due to the sorption of calcium by organic matter, the presence of colloidal gypsum particles or the presence of other calcium- and/or sulfate-containing mineral colloidal particles (van Den Ende, 1991).

4.4.2 Biological sulfate reduction for the treatment of gypsiferous soils (GMOB)

The GMOB contained 25% gypsum prior to treatment. Soils containing 25% gypsum are normally considered to be unsuitable for the growth of most crop types (Verheye & Boyadgiev, 1997). The lowest sulfate concentration of the treated GMOB was 42 g kg^{-1} in a 40% organic mixture. This corresponds to the reduction of the gypsum content from 25% (GMOB) to 7.5% (Figure 4.4). However, the calcium and sulfate content in the treated GMOB still can be categorized as a rather very high level for some sensitive crops (Verheye & Boyadgiev, 1997). This soil may be suitable for moderately tolerant and tolerant crops such as sugar beet, maize, rubber trees, alfalfa, clover and oats (Smith & Robertson, 1962; Verheye & Boyadgiev, 1997). Further studies are nevertheless necessary to explore the agricultural potential of the treated soils.

The pH of the soil mixtures was maintained in the neutral range (pH 7) throughout the experiment without any pH correction (Figure 4.5c). The pH of all soil mixtures increased as the experiment progressed, except for the control unit where the pH was constant at 7.0 since the startup period. The increase of the soil pH was due to alkalinity generation during sulfate reduction, as per the following equation (Sawyer et al., 2003):

$$SO_4^{2-} + Organic \quad Matter \rightarrow HS^- + HCO_3^- \tag{4.1}$$

The pH of soils supplemented with organic substrates deceased during the start-up period (Figure 4.5c). This may be because during this period, the sulfate reduction still did not start, whereas the degradation of the organic substrates to volatile fatty acids (VFAs), such as acetic, propionic and butyric acid by acidifying bacteria (Liamleam & Annachhatre, 2007), resulted in the acidification of the soil.

The produced sulfide was lower than the stoichiometric amount of the reduced sulfate, probably due to binding with metals in the soil or volatilization of H$_2$S in the system. Other researchers have also reported pH dependent loss of sulfur by volatilization of H$_2$S (Jong & Parry, 2003). Moreover, the generated sulfide may precipitate with heavy metals in the soil (Kijjanapanich et al., 2013). The sulfide produced during this biological process can also be used for heavy metal removal from AMD in the mining areas itself as well as from other wastewaters (Jong & Parry, 2003; Kijjanapanich et al., 2012; Liamleam, 2007). Alternatively, it can be used for recovery of elemental sulfur (S^0) (Dutta et al., 2008) or sulfuric acid (H$_2$SO$_4$) (Laursen & Karavanov, 2006).

Most of the gypsiferous soils have a relatively low organic matter content (Ghabour et al., 2008), as also the GMOB used in this study (Table 4.2). Therefore, external electron donor needs to be added for the SRB when these soils are treated by biological sulfate reduction. An electron donor is the primary requirement for SRB (Rzeczycka & Blaszczyk, 2005). Organic wastes are an interesting option as electron donor for SRB, due to their availability, ease for soil application and economic considerations (Costa et al., 2007). The mixtures of PWTS + RH + CHC resulted in conditions that favor the activity and growth of SRB leading to biological sulfate reduction. During the treatment, the GMOB mixture became black and darker as a function of time. This is an indication of the growth and activity of SRB as well as the formation of FeS precipitates. Indeed, blackening of the growth medium is used in diagnostic tests to detect the presence of SRB (Costa et al., 2007).

GMOB before treatment had a very low OM and total nitrogen content. Mixing with the organic mixture resulted in very high levels of OM, total nitrogen, phosphorus and potassium in the GMOB, which remained at a high level after the treatment (Table 4.4). This was due to the nutrients and organic matter contained in the organic substrates, which is another advantage of using organic substrates as electron donor in the SRB based bioremediation technique.

The remediation of gypsiferous soils by biological sulfate reducing process can be applied for either *ex situ* or *in situ* gypsiferous soils treatment. In practice, direct recovery of sulfur from the gas phase may be complicated and difficult, especially in case of the *in situ* treatment which normally covers enormous areas. Therefore, recovery of sulfur from sulfide contained in the leachate of the system can be an alternative option. Sulfide oxidation for elemental sulfur recovery can be done by either abiotic, such as chemical oxidation and electrochemical techniques (Dutta et al., 2008), or by biological oxidation (Sahinkaya et al., 2011). Further studies of treating gypsiferous soils from different sources using this biological sulfate reduction system are recommended to compare and investigate the effect of the soil composition on the sulfate reduction process.

4.5. Conclusions

- Mixtures of PWTS + RH + CHC developed conditions that stimulate the activity and growth of SRB, leading to biological sulfate reduction in gypsiferous soils (GMOB).
- The highest sulfate removal efficiency of 59% was achieved when the soil was mixed with 40% of the organic mixture. This corresponds to the reduction of the gypsum content of the soil from 25 to 7.5%.
- Mixtures of no or low cost organic substrates, such as PWTS + RH + CHC, can be utilized as electron donor for growth of SRB for the removal of sulfate from gypsiferous soils when applying soil bioremediation.

4.6 References

Acar, Y. B., Gale, R. J., Alshawabkeh, A. N., Marks, R. E., Puppala, S., Bricka, M., et al. (1995). Electrokinetic remediation: Basics and technology status. *J. Hazard. Mater., 40*, 117-137.

Alfaya, F., Cuenca-Sánchez, M., Garcia-Orenes, F., & Lens, P. N. L. (2009). Endogenous and bioaugmented sulphate reduction in calcareous gypsiferous soils. *Environ. Technol., 30*(12), 1305-1312.

Battistoni, P., Carniani, E., Fatone, F., Balboni, P., & Tornabuoni, P. (2006). Phosphogypsum leachate: Treatment feasibility in a membrane plant. *Ind. Eng. Chem. Res., 45*, 6504-6511.

Bray, R. H., & Kurtz, L. T. (1945). Determination of total, organic, and available forms of phosphorus in soils. *Soil Sci., 59*(39-45).

CA WET. (1984). California Waste Extraction Test (CA WET) *California Code of Regulations* (pp. 5 (Appendix II)).

Costa, M. C., Martins, M., Jesus, C., & Duarte, J. C. (2007). Treatment of acid mine drainage by sulfate-reducing bacteria using low cost matrices. *Water Air Soil Pollut., 189*(1-4), 149-162.

Dent, D. L. (1986). *Acid sulphate soils: A baseline for research and development* (Vol. 39). Wageningen: International Institute for Land Reclamation and Improvement.

Dutta, P. K., Rabaey, K., Yuan, Z., & Keller, J. (2008). Spontaneous electrochemical removal of aqueous sulfide. *Water Res., 42*, 4965-4975.

Eaton, A. D., APHA, AWWA, & WEF. (2005). *Standard methods for the examination of water and wastewater* (21st ed.). Washington D.C.

FAO. (1974). *FAO-UNESCO Soil map of the world* (Vol. 1). Paris: UNESCO.

FAO. (1990). *FAO Soils Bulletin 62: Management of gypsiferous soils*. Rome.

Ghabour, T. K., Aziz, A. M., & Rahim, I. S. (2008). Anthropogenic impact of fertilization on gypsiferous soils. *Am. Eurasian J. Agric. Environ. Sci., 4*(4), 405-409.

Jang, Y. C., & Townsend, T. (2001). Sulfate leaching from recovered construction and demolition debris fines. *Adv. Environ. Res., 5*, 203-217.

Johnson, D. B., & Hallberg, K. B. (2005). Acid mine drainage remediation options: a review. *Sci. Total Environ., 338*, 3-14.

Jong, T., & Parry, D. L. (2003). Removal of sulfate and heavy metals by sulfate-reducing bacteria in short term bench scale upflow anaerobic packed bed reactor runs. *Water Res., 37*, 3379-3389.

Khresat, S. A., Rawajfih, Z., Buck, B., & Monger, H. C. (2004). Geomorphic features and soil formation of arid lands in Northeastern Jordan. *Arch. Agron. Soil Sci., 50*, 607-615.

Kijjanapanich, P., Annachhatre, A. P., & Lens, P. N. L. (2013). Biological sulfate reduction for treatment of gypsum contaminated soils, sediments and solid wastes. *Crit. Rev. Environ. Sci. Technol., In Press*.

Kijjanapanich, P., Pakdeerattanamint, K., Lens, P. N. L., & Annachhatre, A. P. (2012). Organic substrates as electron donors in permeable reactive barriers for removal of heavy metals from acid mine drainage. *Environ. Technol., 33*(23), 2635-2644.

Kordlaghari, M. P., & Rowell, D. L. (2006). The role of gypsum in the reactions of phosphate with soils. *Geoderma, 132*, 105-115.

Laursen, J. K., & Karavanov, A. N. (2006). Processes for sulfur recovery, regeneration of spent acid, and reduction of NO_x emissions. *Chem. Pet. Eng., 42*(5-6), 229-234.

Liamleam, W. (2007). *Zinc removal from industrial discharge using thermophilic biological sulfate reduction with molasses as electron donor*. Asian Institute of Technology, Thailand.

Liamleam, W., & Annachhatre, A. P. (2007). Electron donors for biological sulfate reduction. *Biotechnol. Adv., 25*(5), 452-463.

Moutsatsou, A., Gregou, M., Matsas, D., & Protonotarios, V. (2006). Washing as a remediation technology applicable in soils heavily polluted by mining-metallurgical activities. *Chemosphere, 63*, 1632-1640.

Pansu, M., & Gautheyrou, J. (2006). *Handbook of Soil Analysis*: Springer.

Pollution Control Department of Thailand. (1994). Surface water quality standard of Thailand. Retrieved 17-09, 2012, from http://www.pcd.go.th/

Rzeczycka, M., & Blaszczyk, M. (2005). Growth and activity of sulphate-reducing bacteria in media containing phosphogypsum and different sources of carbon. *Pol. J. Environ. Stud., 14*(6), 891-895.

Sahinkaya, E., Hasar, H., Kaksonen, A. H., & Rittmamm, B. E. (2011). Performance of a sulfide-oxidizing, sulfur-producing membrane biofilm reactor treating sulfide-containing bioreactor effluent. *Environ. Sci. Technol., 45*(9), 4080-4087.

Sawyer, C. N., McCarty, P. L., & Parkin, G. F. (2003). *Chemistry for environmental engineering and science* (5th ed.): Mc Graw-Hill International

Shternina, E. B. (1960). Solubility of gypsum in aqueous solutions of salts. *Int. Geol. Rev., 1*, 605-616.

Smith, R., & Robertson, V. S. (1962). Soil irrigation classification of shallow soils overlying gypsum beds, northern Iraq. *J. Soil Sci., 13*, 106-115.

U.S.EPA. (1994). METHOD 1312: Synthetic Precipitation Leaching Procedure. Retrieved from http://www.epa.gov/osw/hazard/testmethods/sw846/pdfs/1312.pdf

USDA. (1975). *Soil Taxonomy: A basic system of soil classification for making and interpreting soil surveys*: USDA Soil Conservation Service, Agric. Handbook 436

van Den Ende, J. (1991). Supersaturation of soil solutions with respect to gypsum. *Plant and Soil, 133*, 65-74.

Verheye, W. H., & Boyadgiev, T. G. (1997). Evaluating the land use potential of gypsiferous soils from field pedogenic characteristics. *Soil Use Manage., 13*, 97-103.

Virkutyte, J., Sillanpää, M., & Latostenmaa, P. (2002). Electrokinetic soil remediation-critical overview. *Sci. Total Environ., 289*, 97-121.

Walkley, A., & Black, I. A. (1934). An examination of the Degtjareff method for determining organic carbon in soils: Effect of variations in digestion conditions and of inorganic soil constituents. *Soil Sci., 63*, 251-263.

CHAPTER 5

Biological Sulfate Removal from Gypsum Contaminated Construction and Demolition Debris

This chapter has been published as:
Kijjanapanich, P., Annachhatre, A. P., Esposito, G., van Hullebusch, E. D., & Lens, P.
N. L. (2013). Biological sulfate removal from gypsum contaminated construction and
demolition debris. *J. Environ. Manage., 131*, 82-91.
DOI: 10.1016/j.jenvman.2013.09.025

Chapter 5

Construction and demolition debris (CDD) contains high levels of sulfate that can cause detrimental environmental impacts when disposed without adequate treatment. In landfills, sulfate can be converted to hydrogen sulfide under anaerobic conditions. CDD can thus cause health impacts or odor problems to landfill employees and surrounding residents. Reduction of the sulfate content of CDD is an option to overcome these problems. This study aimed at developing a biological sulfate removal system to reduce the sulfate content of gypsum contaminated CDD in order to decrease the amount of solid waste, to improve the quality of CDD waste for recycling purposes and to recover sulfur from CDD. The treatment leached out the gypsum contained in CDD by water in a leaching column. The sulfate loaded leachate was then treated in a biological sulfate reducing Upflow Anaerobic Sludge Blanket (UASB) reactor to convert the sulfate to sulfide. The UASB reactor was operated at $23 \pm 3°C$ with a hydraulic retention time and upflow velocity of 15.5 h and 0.1 m h^{-1}, respectively while ethanol was added as electron donor at a final organic loading rate (OLR) of 3.46 g COD L^{-1} reactor d^{-1}. The CDD leachate had a pH of 8-9 and sulfate dissolution rates of 526.4 and 609.8 mg L^{-1} d^{-1} were achieved in CDD gypsum and CDD sand, respectively. Besides, it was observed that the gypsum dissolution was the rate limiting step for the biological treatment of CDD. The sulfate removal efficiency of the system stabilized at around 85%, enabling the reuse of the UASB effluent for the leaching step, proving the versatility of the bioreactor for practical applications.

5.1. Introduction

Gypsum is mined, processed and converted into several products; it is widely used in the construction industry and it is a major component in drywalls (gypsum boards). Construction, renovation or demolition activities yield large amounts of wastes called construction and demolition debris (CDD). A typical CDD composition includes wood, concrete, rock, paper, plastic, gypsum drywall and heavy metals (Table 2.2). It has been reported that, on an average, 0.9 metric tons of gypsum waste is generated from the construction of a typical single family home or 4.9 kg of waste gypsum is generated per square meter of the building structure (Turley, 1998). According to the U.S. EPA characterization studies of CDD, gypsum drywall accounts for 21-27% of the mass of debris generated during the construction and renovation of residential structures (U.S.EPA, 1998). Nearly 40% of the total mass of CDD consists of a fine fraction containing high amounts of gypsum (Montero et al., 2010; Townsend et al., 2004), namely CDD sands (CDDS). The gypsum content (mass basis) in CDDS ranges from 1.5 to 9.1% (Jang & Townsend, 2001a).

Reuse options have been proposed for CDD, including soil amendment, alternative daily landfill cover, or fill material for the construction of roads, embankment and other construction projects (Jang & Townsend, 2001a). The presence of gypsum drywall in CDD may provide some benefits, depending on the application, e.g. as soil conditioner or nutrient source. However, for applications where the material is placed in direct contact with the environment, there are potential regulatory concerns regarding the high levels of sulfate and heavy metals in CDD and the potential risks to human health and the environment (Jang & Townsend, 2001a). The Dutch government has set the limits to the maximum amount of polluting compounds present in building material. For reusable sand, the emission limit is 1.73 g sulfate per kg of sand (de Vries, 2006; Stevens, 2013).

Therefore, most of the CDD cannot be reused for construction activities due to its high sulfate content. Moreover, deposition of CDD in landfills can lead to exceptionally high levels of biogenic sulfide formation (H_2S), posing serious problems such as odor (Jang, 2000; Lens & Kuenen, 2001), pipe corrosion (Vincke et al., 2001) and contamination of landfill gas (Karnachuk et al., 2002) or groundwater. Thus, gypsum waste has to be separated from other wastes, especially organic waste, and placed in a specific area of a landfill. This results in a rise of the disposal costs of gypsum waste (Gypsum Association, 1992).

Montero et al. (2010) showed that organic matter, such as wood and paper, in CDD is distributed mainly in the fractions of large-sized components, while the gypsum is concentrated mostly in the fine fractions (52.4% of total sulfate). As a result, the amount of gypsum to be disposed to the landfill can be reduced by separating the fine fraction from the mixed CDD. However, final disposal still requires removing gypsum from the fine fraction.

This study aimed at developing a biological sulfate removal system (Figure 5.1) to reduce the sulfate content of CDD and recover the sulfur from the solid waste, which not only decreases the amount of solid waste but also improves the quality of wastes (CDD and sulfur) making it suitable for recycling purposes. In particular, this research investigated the leachability of CDD gypsum in a leaching column, and if the sulfate containing leachate could be further treated in a biological sulfate removal step in order to reuse the UASB effluent in the leaching column to leach the CDD. The sulfate removal step utilized the bacterial sulfate reduction process as it occurs in nature for the conversion of sulfate to sulfide.

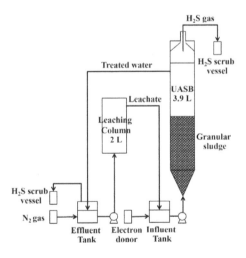

Figure 5.1. Schematic representation of the biological CDD treatment system: CDDG or CDDS leaching column coupled to a UASB reactor for biological sulfate removal.

5.2. Material and Methods
5.2.1 Construction and demolition debris (CDD)

CDD samples were collected from Smink Afvalverwerking B.V. (Amersfoort, The Netherlands). The samples were air-dried and divided into two parts by sieving at 2 mm. Pieces of wood, concrete, rock, paper, plastic and foam were removed by hand (particles larger than 2 mm) retaining only drywall particles, namely CDD gypsum (CDDG). The materials with a particle size smaller than 2 mm also contained sand fraction, and was called CDD sand (CDDS). The gypsum content of the CDDG and CDDS were 37 and 16% w/w, respectively.

5.2.2 Sulfate reducing bacteria (SRB) inoculums

A mixture of anaerobic granular sludges from UASB and Expanded Granular Sludge Bed (EGSB) systems treating pulp & paper and food industrial wastewater, provided by Biothane Systems International (Delft, The Netherlands), was used as a source of sulfate reducing bacteria (SRB). The seed sludge had a total suspended solids (TSS) and volatile suspended solids (VSS) content of 54.6 g L^{-1} and 39.8 g L^{-1}, respectively, corresponding to a VSS/TSS ratio of 0.73.

5.2.3 Leaching of gypsum in batch experiments

The batch experiments on leaching of gypsum were carried out both on CDDG and CDDS. CDDG and CDDS were washed with demineralized water using 1:10 ratio of CDD: demineralized water and then placed on a rotary shaker for 24 h at 150 rpm. The supernatant was filtered and analyzed at the end of the experiment for pH, sulfate, metals and some macro nutrients. The experiments were conducted at room temperature (23 ± 3°C).

5.2.4 Leaching of gypsum in continuous column experiments

The leaching columns were made of polyvinyl chloride (PVC) which had a working volume of 2 L. One kg of CDDG or CDDS was filled in each leaching column. Demineralized water was fed at the bottom of the column at a flow rate of 252 mL h^{-1} (upflow velocity 0.1 m h^{-1}) using a peristaltic pump. The leachate was withdrawn from the top of the column and collected daily for analysis. The experiments were conducted at room temperature (23 ± 3°C) and lasted for 20 and 60 d for CDDS and CDDG, respectively.

5.2.5 Leachate treatment in bioreactor experiments

The combined system was divided into two parts: a leaching column coupled to a UASB reactor (Figure 5.1). The effluent of the UASB reactor was reused in the leaching column. No pH adjustment was done in any of the systems. The leaching columns had similar details as those described above. The leaching column was filled with CDD containing about 0.5 kg of CDDG or 1 kg of CDDS (equivalent to 100 g of sulfate). The leachate from the leaching column was supplemented with ethanol (OLR of 1.75-3.46 g COD L^{-1} reactor d^{-1}) and fed to the UASB reactor. The effluent of the UASB reactor was withdrawn from the top of the reactor and reused as leaching water.

The UASB reactor was made of polymethyl-methacrylate (PMMA) and had a working volume of 3.9 L (Figure 5.1). The UASB reactor was inoculated with anaerobic granular sludge (50% by volume) and operated at room temperature ($23 \pm 3°C$). The influent (CDDG or CDDS leachate) was fed using a peristaltic pump at the bottom of the UASB reactor at a flow rate of 252 mL h^{-1}, resulting in a hydraulic retention time (HRT) and liquid upflow velocity of 15.5 h and 0.1 m h^{-1}, respectively.

The treatment of the CDDG leachate in the UASB reactor was investigated at two OLRs. In the first experiment (UASB I), ethanol was fed to the reactor influent at 480 mg L^{-1}, corresponding to an OLR of 1.75 g COD L^{-1} reactor d^{-1}. The OLR was increased to 3.46 g COD L^{-1} reactor d^{-1} (ethanol 950 mg L^{-1}) on day 24. This OLR was maintained until the end of the experiment. A second experiment (UASB II) on the leaching- UASB system treating CDDG was operated at an OLR of 3.46 g COD L^{-1} reactor d^{-1}. This OLR (3.46 g COD L^{-1} reactor d^{-1}) was also applied to the entire experimental run of the UASB reactor treating the CDDS leachate (UASB III). In UASB II and III, the treated water was purged with nitrogen gas (N_2) (at a 10 L h^{-1} flow rate) to remove H_2S prior to recycling the UASB effluent to the leaching column. This resulted in a sulfide concentration of around 20 mg S L^{-1} in the influent supplied to the leaching column, for UASB II and III experiments, respectively.

5.2.6 Analytical methods

The evolution of the leachate color was measured via the absorbance at 200-800 nm using a Perkin Elmer Lambda 20 UV visible spectrophotometer. The pH was measured using a 691 Metrohm pH meter and a SenTix 21 WTW pH electrode, while the oxidation-reduction potential (ORP) was measured using a 340i WTW pH meter and a QR481X QIS ORP electrode. The Electro-Conductivity (EC) was measured using a LF 323 WTW conductivity meter.

Sulfate was measured using an ICS-1000 Dionex Ion Chromatography (IC) (Eaton et al., 2005). Sulfide was measured by the method proposed by Ralf Cord-Ruwisch (Ruwisch, 1985) using a Perkin Elmer Lambda 20 UV visible spectrophotometer. Calcium was measured using an AAnalyst 200 Perkin Elmer Atomic Absorption Spectrometer (AAS)-Flame (Eaton et al., 2005). Metals and some macro nutrients (Na, Mg and K) were measured using Thermo Scientific XSeries 2 inductively coupled plasma mass spectrophotometer (ICP-MS). Ethanol and acetate were measured using a Varian 430 Gas Chromatograph (GC) (Eaton et al., 2005). The dissolved organic carbon (DOC) was monitored as an indicator of dissolved carbon available for bacterial metabolism. DOC was measured using the high temperature combustion method by the Shimadzu TOC-V CPN analyzer (Eaton et al., 2005).

X-ray diffraction (XRD) analysis was performed on a Bruker D8 Advance diffractometer equipped with an energy dispersion Sol-X detector with copper radiation (CuKα, $\lambda = 0.15406$ nm). The acquisition was recorded between 2° and 80°, with a 0.02° scan step and 1 s step time. Samples were previously dried at 25°C and crushed prior to XRD analysis.

5.3. Results
5.3.1 Leaching experiments
5.3.1.1 Batch experiments

Table 5.1 shows the characteristics of the CDDG and CDDS leachate. It was observed that most of the metals and their respective concentration in CDDS leachate were higher than in the CDDG leachate. A large variation of the metal concentration was found in the case of CDDG leachate.

Table 5.1. Characteristics of CDDG and CDDS leachate (Solid:Liquid ratio = 1:10)

Parameter		CDDG	CDDS
pH		7.59 ± 0.09	7.71 ± 0.07
Macro Nutrients (mg L^{-1})	Sulfate	1662.66 ± 33.38	1760.24 ± 30.44
	Na	29.92 ± 2.77	40.85 ± 0.67
	Mg	6.47 ± 2.32	21.86 ± 1.51
	K	20.80 ± 1.60	27.58 ± 0.59
	Ca	570.67 ± 6.11	589.33 ± 8.33
Heavy Metals (µg L^{-1})	Al	68.83 ± 7.58	57.64 ± 17.18
	Cr	2.60 ± 0.52	<2
	Mn	47.00 ± 22.56	199.93 ± 6.31
	Fe	44.23 ± 17.58	81.41 ± 15.93
	Co	4.50 ± 2.59	5.60 ± 0.15
	Ni	6.69 ± 0.98	11.32 ± 0.37
	Cu	18.08 ± 2.82	30.68 ± 1.65
	Zn	34.87 ± 32.86	67.58 ± 2.38
	As	4.65 ± 0.76	5.22 ± 0.08
	Mo	14.27 ± 10.13	26.54 ± 2.21
	Cd	<2	<2
	Ba	69.37 ± 1.33	71.80 ± 0.90
	Pb	45.47 ± 67.34	6.12 ± 1.53

5.3.1.2 Column experiments

The DOC of the leachate from the CDDG and CDDS decreased from 44 to 2 mg L^{-1} after 5 d and from 133 to 3 mg L^{-1} after 10 d, respectively (Figure 5.2a). The leachate samples from CDDG and CDDS had a yellow color, which diminished as time progressed (Figures 5.2b and 5.2c).

The pH of the leachate from CDDG and CDDS increased from 7 to 8 within 2 and 8 d, respectively. Then, the pH of both leachates remained at around 8-9 throughout the experiments (Figure 5.3a). The EC of the CDDG leachate decreased at a moderate pace, while the EC of the CDDS decreased rapidly (Figure 5.3b).

The sulfate concentration of the CDDG leachate decreased moderately from 1400 to 10 mg L^{-1}, while the sulfate concentration decreased rapidly from 1900 to 7 mg L^{-1} in case of the CDDS leachate (Figure 5.3c). Average sulfate dissolution rates of 526.4 and 609.8 mg L^{-1} d^{-1} were achieved in CDDG and CDDS, respectively. Around 200 g and 95 g of sulfate were removed from 1 kg of CDDG and CDDS during 60 and 20 d, respectively (Figure 5.3d).

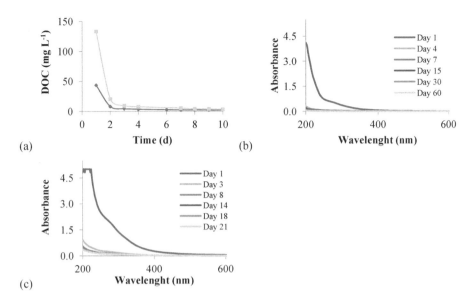

Figure 5.2. Evolution of leachate DOC and color over leaching time: (a) leachate DOC, (b) CDDG leachate absorbance at 200-800 nm and (c) CDDS leachate absorbance at 200-800 nm. (♦) CDDG and (■) CDDS.

The dissolution of calcium from the CDDG and CDDS followed a pattern similar to the sulfate dissolution. Average calcium dissolution rates of 240.6 and 221.8 mg L^{-1} d^{-1} were achieved in the CDDG and CDDS columns, respectively (Figure 5.3e). Around 90 g and 35 g of calcium were removed from 1 kg of CDDG and CDDS, respectively (Figure 5.3f). The Ca^{2+}/SO_4^{2-} ratio of CDDG and CDDS was 1.1 and 0.9 (mole basis), respectively, which was not significantly different from the theoretical ratio of gypsum (Ca^{2+}/SO_4^{2-} ratio = 1). In addition, the mass loss from the CDDG and CDDS after the leaching experiment was 39 and 16% w/w, respectively, which was almost equal to the amount of gypsum removed (37 and 16% w/w, respectively).

The CDDG and CDDS samples in columns before and after the leaching step were analyzed by XRD. From Figure 5.4, it is clearly evident that crystalline gypsum was not present in both CDDG and CDDS after the leaching experiment.

5.3.2 Bioreactor experiments
5.3.2.1 Bioreactors treating CDDG leachate (UASB I and UASB II)

The UASB I and II influent and effluent pH values in both experiments remained in the neutral range (pH 6-8) without pH adjustment (data not shown). The ORP of the UASB I and II effluents stabilized in the range between -375 and -391 mV throughout the experimental period.

In UASB I, there was a significant fluctuation in the influent and effluent sulfide concentration (Figure 5.5a). The effluent sulfate concentration decreased as the experiment progressed (Figure 5.5c). However, this reduction in sulfate concentration was slow (386 mg L^{-1} d^{-1}) and the sulfate removal efficiency was around 25-45% only.

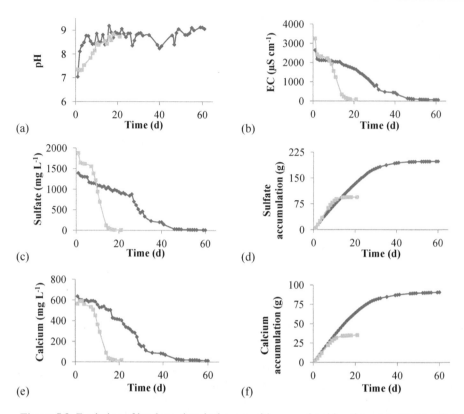

Figure 5.3. Evolution of leachate chemical composition over leaching time: (a) pH, (b) EC, (c) sulfate, (d) sulfate accumulation, (e) calcium and (f) calcium accumulation. (♦) CDDG and (■) CDDS.

After increasing the OLR to 3.46 g COD L^{-1} reactor d^{-1}, there was a rapid increase in the sulfate removal efficiency, which reached more than 95% within 3 d (Figure 5.5c). The sulfate concentration of the effluent significantly decreased to below 20 mg L^{-1} within 4 d. Calcium concentrations in the influent varied between 590 and 690 mg L^{-1} which further increased up to 800 mg L^{-1} after increasing the OLR (Figure 5.5e).

While no ethanol was detected in both (UASB I and UASB II) effluents (data not shown), acetate was measured up to 1620 mg L^{-1} (Figures 5.5g and 5.5h). In UASB I, the DOC removal efficiency was up to 96% (Figure 5.5i). When the OLR was increased to 3.46 g COD L^{-1} reactor d^{-1} (influent ethanol concentration 950 mg L^{-1}), the DOC removal efficiency decreased to 50% and acetate started to accumulate within the bioreactor. On day 15, there was a rapid drop in the DOC removal efficiency of UASB I caused by clogging of the sludge bed (Figure 5.5i). This was presumably due to the precipitation of calcium carbonate onto the granular sludge surface (data not shown). Therefore, the system was stopped for maintenance on that day. In UASB II, the DOC removal efficiency remained around 55% throughout the experiment (Figure 5.5j).

In UASB II and UASB III, the treated water was purged with nitrogen gas (N_2) (at a 10 L h^{-1} flow rate) to remove H_2S prior to recycling the UASB effluent to the leaching column. This resulted in a sulfide concentration of around 20 mg S L^{-1} in the influent supplied to the leaching column. The average UASB II effluent sulfide concentration was 130 mg S L^{-1} (Figure 5.5b). The sulfate removal efficiency of UASB II was around 50% during the first 20 d of the experiment and this value gradually increased up to 95% at the end of the experiment (day 35, Figure 5.5d). The calcium concentration in the UASB II influent and effluent was not significantly different with the average around 677 mg L^{-1} (Figure 5.5f). The XRD results show that some gypsum still remained in CDDG upon termination of the experiment (Figure 5.6a).

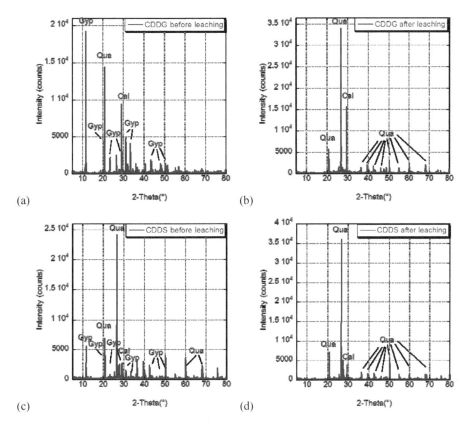

(a) (b) (c) (d)

Figure 5.4. XRD spectra of (a) CDDG before leaching, (b) CDDG after leaching, (c) CDDS before leaching and (d) CDDS after leaching. (Qua) Quartz, (Gyp) Gypsum, (S) Elemental sulfur and (Cal) Calcite.

5.3.2.2 Bioreactor treating CDDS leachate (UASB III)

The UASB III influent and effluent pH values remained in the neutral range (pH 6-8) without pH adjustment (Figure 5.7a). There was a fluctuation in the influent and effluent sulfide concentration and it was observed that the sulfide production was lower than the sulfate reduction (Figure 5.7b). The ORP of the UASB III effluent stabilized in the range between -380 and -393 mV throughout the experiment.

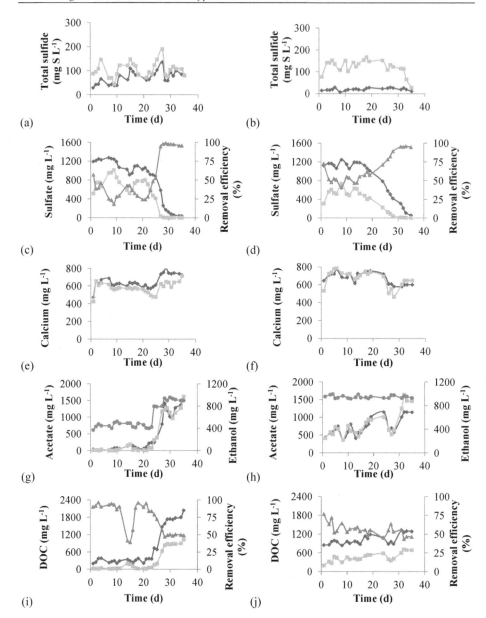

Figure 5.5. Performance of a bioreactor treating CDDG as a function of operation time (Left: UASB I and Right: UASB II): (a, b) total sulfide, (c, d) sulfate, (e, f) calcium, (g, h) ethanol and acetate and (i, j) DOC. (♦) influent, (■) effluent, (▲) removal efficiency and (●) influent ethanol.

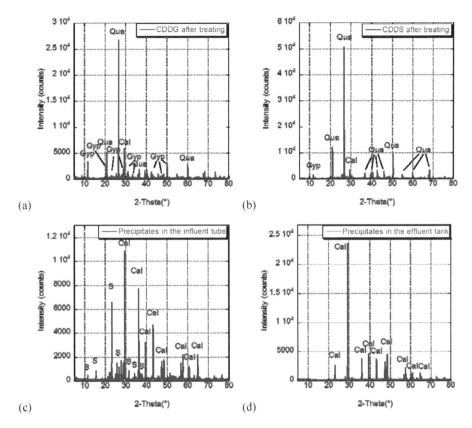

Figure 5.6. XRD spectra of (a) CDDG after treatment, (b) CDDS after treatment, (c) precipitate in the UASB influent tube and (d) precipitate in the effluent tank. (Qua) Quartz, (Gyp) Gypsum, (S) Elemental sulfur and (Cal) Calcite.

The sulfate concentration of the UASB III effluent decreased rapidly within the first 6 d, while the sulfate removal efficiency increased rapidly (Figure 5.7c). After the first 6 d of the experiment, the sulfate removal efficiency of the system stabilized at around 85%. Figure 5.6b shows that there was gypsum left in the CDDS upon termination of the treatment. A white-yellow precipitate was found in the influent tube of both the CDDG and CDDS treatment systems. This precipitate was composed of elemental sulfur (S^0) and calcite (Figure 5.6c).

Ethanol (950 mg L^{-1}) was supplied to UASB III at an OLR of 3.46 g COD L^{-1} reactor d^{-1}. Although, ethanol was not detected in the UASB III effluent (data not shown), the acetate concentration in the effluent was found to be around 1000 mg L^{-1} (Figure 5.7e), and the average DOC removal efficiency was only 55% (Figure 5.7f).

Concerning calcium concentrations, no significant difference between the UASB III influent and effluent values was noticed (Figure 5.7d). The calcium concentration increased slightly from 540 to 760 mg L^{-1} during the first 16 d, and then dropped to around 500 mg L^{-1} during the end of the experiment. After day 16, a precipitate was observed in the effluent tank, which was identified to be calcite (Figure 5.6d).

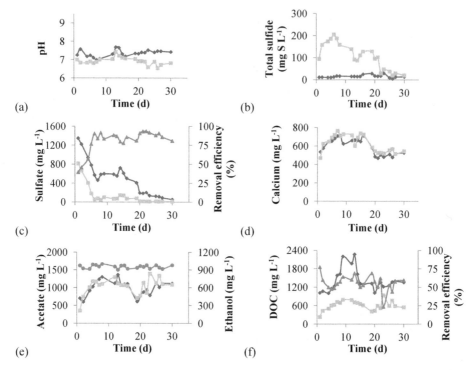

Figure 5.7. Performance of a bioreactor treating CDDS (UASB III) as a function of operation time: (a) pH, (b) total sulfide, (c) sulfate, (d) calcium (e) ethanol and acetate and (f) DOC. (♦) influent, (■) effluent, (▲) removal efficiency and (●) influent ethanol.

5.4. Discussion

5.4.1 Leaching of construction and demolition debris (CDD)

This study showed that the sulfate content removed in the column experiment from the CDDS used in this research (95 g kg^{-1}) is much higher than the sulfate content in the CDDS reported in the study of de Vries (2006) (25 g kg^{-1}), who also tested CDDS from the same company. Such variation in the sulfate content was also observed in the study of Jang and Townsend (2001a) where the sulfate content of CDDSs collected from 13 CDD recycling facilities in south Florida varied between 8.4 and 51.0 g kg^{-1} CDDS. The characteristics of CDD varies depending on the source of the CDD (Townsend et al., 2004).

The pH of both the CDDG and CDDS leachate was around 7.6 and 7.7, respectively (batch experiments), while in column experiments increased to around 8.0-9.0 after the start-up phase (Figure 5.3a). This observation is in agreement with de Vries (2006) and Jang and Townsend (2001a), where the pH of the leachate of the CDDS increased from neutral at the start to 9.2 and 10.4, respectively, at the end of their experiment (7 and 90 d, respectively). Such a pH rise to above 7 is due to the dissolution of sulfate, resulting in a pH increase (de Vries, 2006). However, several other components present in CDD can contribute to the alkalinity of the leachate, such as cement, concrete dust (Townsend et al., 2004) or calcite (Coto et al., 2012) .

The solubility of gypsum varies as a function of temperature, particle size and the presence of other salts in the system (Verheye & Boyadgiev, 1997). Gypsum has a solubility of 2600 mg L^{-1} in pure water at 25°C (FAO, 1990), which results in a sulfate concentration of 1450 mg L^{-1}. The maximum solubility (2720 mg L^{-1} or 20 mM) of gypsum occurs at 30-40°C (van Driessche et al., 2012; Verheye & Boyadgiev, 1997). From Figure 5.3, CDDS, which has a smaller particle size, has a higher sulfate dissolution rate (609.8 mg L^{-1} d^{-1}) than CDDG (526.4 mg L^{-1} d^{-1}). The highest sulfate concentration found in this study was 1876 mg L^{-1} from the CDDS leachate at 23 ± 3°C (Figure 5.3c). The sulfate concentration of the leachate which exceeded the solubility limit indicates that the leachate samples were supersaturated with sulfate. Jang and Townsend (2001a) found that the concentrations of sulfate of many leachate samples exceed the solubility limit from gypsum dissolution (up to 1585 mg L^{-1}). This is possible due to the presence of other ions such as sodium and chloride (data not shown) and the increased ionic strength of the leachate (Jang & Townsend, 2001a). For instance, the gypsum solubility was found to be 3 times higher in the presence of 5 g L^{-1} sodium (Shternina, 1960). In this study, the highest sulfate concentration (1876 mg L^{-1}) was found in CDDS leachate, which contains sodium, magnesium, potassium in a higher concentration than those in CDDG leachate (Table 5.1). Super-saturation of sulfate can also occur due to the complexation of calcium by organic matter, the presence of colloidal gypsum particles or the presence of other calcium- and/or sulfate-containing mineral colloidal particles (van Den Ende, 1991). However, the dissolution rate of gypsum was low (526.4 - 609.8 mg L^{-1} d^{-1}). Therefore, the leaching step is the most time consuming step; hence, further research is recommended to develop appropriate techniques to increase the gypsum dissolution rate, such as increasing the temperature or decreasing the pH of the leachate.

The leachate samples from CDDG and CDDS had a yellow color (maximum absorption at 200 nm) (Figures 5.2b and 5.2c) and contained DOC, suggesting that besides gypsum, other organic compounds were also possibly leached out from the CDD during the initial phase. Due to the low amounts of DOC in the leachate, an external carbon source needs to be supplied to support SRB activity and growth. Jang and Townsend (2001b) analyzed volatile and semi-volatile organic compounds present in CDD from 14 CDD recycling facilities in south Florida. They found that toluene showed the highest leachability among the compounds (61.3-92.0%), while trichlorofluoromethane, the most commonly detected compound in CDD, had the lowest leachability (1.4-39.9%). Several polycyclic aromatic hydrocarbons (PAHs) also leached during the leaching tests from CDD waste (Jang & Townsend, 2001b). The results from their study indicate that the organics in CDD recycling facilities were not a major concern, especially from the view-point of human health risk and leaching risk to groundwater under reuse and contact scenarios. However, further research is needed to optimize the PAHs removal efficiency of the UASB to prevent their accumulation and toxicity.

There was no crystalline gypsum left in both CDDG and CDDS after the leaching experiment (Figure 5.4). Moreover, the mass loss from the CDDG and CDDS after the leaching experiment was 39 and 16% w/w, respectively, which was almost equal to the amount of gypsum removed (37 and 16% w/w, respectively). Likewise, the general composition of CDD waste confirms that almost all the gypsum was dissolved into the leachate. For CDDG, another 61% should be concrete or rock which are packed together with gypsum drywall, while 84% should be the remaining sand in case of CDDS. From a practical view-point, CDDG and CDDS after leaching could be reused for construction purpose due to its low sulfate content which is lower than the allowable maximum amount of sulfate present in building sand (1.73 g sulfate kg^{-1} of sand for the Netherlands) (de Vries, 2006; Stevens, 2013).

5.4.2 Treatment of CDD leachate in a sulfate reducing bioreactor

The gypsum contained in the CDD is leached out using water in a leaching column (Figure 5.1). The sulfate containing leachate is further treated in a biological sulfate reducing UASB reactor. Sulfate removal not only solves sulfate contamination problems, but the sulfide produced during this process can also be used for heavy metal removal from the leachate itself as well as from other wastewaters (Jong & Parry, 2003; Kijjanapanich et al., 2012; Liamleam, 2007) or can be recovered as elemental sulfur, while reducing the amount of solid waste that needs to be disposed in the landfill sites. Sulfide which is the product of biological sulfate reduction can be converted either to sulfate or to sulfur during a sulfide oxidation process (Annachhatre & Suktrakoolvait, 2001). Sulfide oxidation for elemental sulfur recovery can be done by abiotic, such as chemical oxidation and electrochemical techniques (Dutta et al., 2008), or by biotic conversions.

For CDDG treatment, the sulfate removal efficiency was low at the first 24 d with an OLR of 1.75 g COD L^{-1} reactor d^{-1}, presumably, due to the toxicity of the accumulated sulfide in the system. In biological sulfate reduction processes, the sulfide generated can be toxic to SRB. H$_2$S is toxic to microorganisms because of its permeability through the cell membrane in its undissociated form (Al-Zuhair et al., 2008; Speece, 1983). Studies where H$_2$S was continuously removed from the SRB growth medium resulted in a 4-5 times higher amount of H$_2$S generated as compared to cells where the H$_2$S was not continuously removed (Gypsum Association, 1992). At a pH below 7.0, H$_2$S is the dominant inhibitor (Al-Zuhair et al., 2008). SRBs are less sensitive to the total sulfide concentration when the pH is increased from 6.8 to 8.0 and more sensitive to the undissociated sulfide (H$_2$S) concentration (O'Flaherty & Colleran, 1999).Therefore, the treated water was purged with nitrogen gas (N$_2$) to remove H$_2$S prior to entering the leaching column in UASB II and III. This process results a higher sulfide production as well as sulfate reduction rates.

The low sulfate removal efficiencies of the CDDG treatment (UASB I) at an OLR of 1.75 g COD L^{-1} reactor d^{-1} might be due to an insufficient addition of electron donor. This can be implied by the increase of the sulfate removal efficiency from 25-45% to 50% when a higher OLR (3.46 g COD L^{-1} reactor d^{-1}) was applied. Although all the electron donor was consumed during the first 24 d (96%), it seemed that insufficient electron was available for the SRB. Indeed, anaerobic granular sludge contains not only SRB but also methane producing organisms and many other bacteria. These can predominate over SRB during the start-up period, and thus consume the supplied electron donor.

Another cause of the low sulfate removal efficiencies of the CDDG treatment might be the presence of impurities, such as heavy metal, contained in CDDG which can inhibit SRB activities (Azabou et al., 2007; Azabou et al., 2005; de Vries, 2006). Apart from xenobiotics (see above), CDD can also be contaminated with heavy metals such as aluminium, arsenic, cadmium, chromium and copper (Townsend et al., 2004). The contamination may come from the soil in the CDD stream itself, from small pieces of hazardous building materials (e.g. paint chips or dust associated with lead-based paint debris) and/or from leaching of hazardous materials commingled with the waste stream (Townsend et al., 2004). These heavy metals present in the gypsum waste can inhibit growth and activity of SRB (Azabou et al., 2005; de Vries, 2006). In this study, the large variation of metal concentration was found in CDDG leachate (Table 5.1). This was due to the large sample sized fractions (2-6 mm) which made it difficult to homogenize. Therefore, high concentration of toxic metal may be found in CDDG leachate. Further study is needed in order to protect and reduce the toxicity of metal present in the CDDG leachate to SRB.

There was a significant fluctuation in the UASB I influent and effluent sulfide concentration (Figure 5.5a). The amount of sulfide produced was lower than the stoichiometric amount of sulfate reduced in all experiments, probably due to volatilization of H_2S in the system. Sulfide can combine with protons to produce H_2S which is pH dependent (Sawyer et al., 2003) and can volatilize as gaseous product. Other researchers have also reported pH dependent loss of sulfur due to volatilization of H_2S (Jong & Parry, 2003). Another reason for lower sulfide production can be partial re-oxidation of sulfide to elemental sulfur. In this study, the precipitates of elemental sulfur were found in the tubing of the system (Figure 5.6c). This may be due to an air leak in the system during the addition of ethanol to the system.

A minor residue of gypsum remained in both the CDDG and CDDS after leaching with the treated water (Figures 5.6a and 5.6b). This may due to insufficient operating time to dissolve all gypsum and the bioreactor could not remove all the sulfate contained in the leachate. However, the levels of remaining gypsum, 0.3-0.7 g sulfate kg^{-1} of sand (calculated from the influent sulfate concentration of the last day of the experiment), is far below the Dutch government limit for the maximum amount of sulfate present in building sand (1.73 g sulfate kg^{-1} of sand) (de Vries, 2006; Stevens, 2013).

All UASB effluents had a high concentration of calcium (590-800 mg L^{-1}), which has to be removed from the leachate prior to discharge into surface water bodies. In this study, calcium carbonate precipitates were present in the UASB effluent tank (Figure 5.6d). Thus, a calcium recovery step might be required to prevent accumulation of calcium carbonate precipitates in the piping or UASB granular sludge, e.g. by microbial carbonate precipitation (MCP) using ureolytic bacteria (Al-Thawadi & Cord-Ruwisch, 2012; Al-Thawadi, 2011; Hammes et al., 2003; Whiffin et al., 2007).

Carbon source and electron donor is the primary substrate required for sulfate reduction using SRB. Ethanol has been used because it is cheap and easy to use. A sulfate conversion efficiency as high as 80% has been reported at high sulfate loading rates (up to 10-12 gS L^{-1} d^{-1}) while using ethanol as electron donor (de Smul et al., 1997). Organic waste can be an interesting alternative, because many companies have such waste streams. A disadvantage is, however, the possible need for a post-treatment step to remove the residual pollution or unwanted waste compounds from the organic waste itself.

When using ethanol as electron donor especially in high rate systems, high concentrations of acetate (up to 1 g L^{-1}) are generated which resulted in an increase in COD of the effluent (Cao et al., 2012; Liamleam & Annachhatre, 2007). Acetate production during the biological sulfate reduction is a major drawback of sulfate reducing reactors because most SRB species present in UASB reactors cannot completely oxidize acetate, even at excess sulfate levels (Lens et al., 2002). Figure 7e shows that acetate can also be consumed by the bacteria present in the system, as the acetate concentration in the effluent was constant at around 1000 mg L^{-1}. However, this concentration is still too high and pollutes the environment when it is discharged without proper treatment. Process control which has been used for several biological production processes can be an alternative option to control the formation of desirable end products in sulfate reduction systems (Dunn et al., 2005; Villa-Gomez et al., 2013). With better process control, excess acetate formation can be avoided, thus decreasing of the operational cost and eliminating the need for a post-treatment step to remove acetate.

5.5. Conclusions

This investigation demonstrated that a water based leaching step coupled to a biological sulfate reduction step can be used for the treatment of CDD, where SRB used the gypsum in the CDD as sulfate source. The sulfate removal efficiency up to 85% was achieved and the levels of remaining gypsum in the treated CDD (0.3-0.7 g sulfate kg^{-1} of sand) is far below the Dutch government limit for the maximum amount of sulfate present in building sand. The developed system was also able to reduce and prevent possible adverse impacts of CDD on the environment.

5.6 References

Al-Thawadi, S., & Cord-Ruwisch, R. (2012). Calcium carbonate crystals formation by ureolytic bacteria isolated from Australian soil and sludge. *J. Adv. Sci. Eng. Res., 2*, 12-26.

Al-Thawadi, S. M. (2011). Ureolytic bacteria and calcium carbonate formation as a mechanism of strength enhancement of sand. *J. Adv. Sci. Eng. Res., 1*, 98-114.

Al-Zuhair, S., El-Naas, M. H., & Al-Hassani, H. (2008). Sulfate inhibition effect on sulfate reducing bacteria. *J. Biochem. Technol., 1*(2), 39-44.

Annachhatre, A. P., & Suktrakoolvait, S. (2001). Biological sulfate reduction using molasses as a carbon source. *Water Environ. Res., 73*, 118-126.

Azabou, S., Mechichi, T., Patel, B. K. C., & Sayadi, S. (2007). Isolation and characterization of a mesophilic heavy-metal-tolerant sulfate-reducing bacterium *Desulfomicrobium* sp. from an enrichment culture using phosphogypsum as a sulfate source. *J. Hazard. Mater., 140*, 264-270.

Azabou, S., Mechichi, T., & Sayadi, S. (2005). Sulfate reduction from phosphogypsum using a mixed culture of sulfate-reducing bacteria. *Int. Biodeter. Biodegr., 56*(4), 236-242.

Cao, J., Zhang, G., Mao, Z., Li, Y., Fang, Z., & Yang, C. (2012). Influence of electron donors on the growth and activity of sulfate-reducing bacteria. *Int. J. Miner. Process., 106-109*, 58-64.

Coto, B., Martos, C., Pena, J. L., Rodríguez, R., & Pastor, G. (2012). Effects in the solubility of CaCO3: Experimental study and model description. *Fluid Phase Equilib. , 324*, 1-7.

de Smul, A., Dries, J., Goethals, L., Grootaerd, H., & Verstraete, W. (1997). High rates of microbial sulphate redction in a mesophilic ethanol-fed expanded-granular-sludge-blanket reactors. *Appl. Microbiol. Biotechnol., 48*, 297-303.

de Vries, E. (2006). *Biological sulfate removal from construction and demolition debris sand.* Wageningen University.

Dunn, I. J., Heinzle, E., Ingham, J., & Přenosil, J. E. (2005). *Biological Reaction Engineering*: Wiley-VCH Verlag GmbH & Co. KGaA.

Dutta, P. K., Rabaey, K., Yuan, Z., & Keller, J. (2008). Spontaneous electrochemical removal of aqueous sulfide. *Water Res., 42*, 4965-4975.

Eaton, A. D., APHA, AWWA, & WEF. (2005). *Standard methods for the examination of water and wastewater* (21st ed.). Washington D.C.

FAO. (1990). *FAO Soils Bulletin 62: Management of gypsiferous soils.* Rome.

Gypsum Association. (1992). Treatment and disposal of gypsum board waste: Technical paper part II, *AWIC's Construction Dimensions* (Vol. March): AWIC.

Hammes, F., Seka, A., de Knijf, S., & Verstraete, W. (2003). A novel approach to calcium removal from calcium-rich industrial wastewater. *Water Res., 37*(3), 699-704.

Jang, Y. (2000). *A study of construction and demolition waste leachate from laboratory landfill simulators.* University of Florida, Florida.

Jang, Y. C., & Townsend, T. (2001a). Sulfate leaching from recovered construction and demolition debris fines. *Adv. Environ. Res., 5*, 203-217.

Jang, Y. C., & Townsend, T. G. (2001b). Occurrence of organic pollutants in recovered soil fined from construction and demolition waste. *Waste Manage., 21*, 703-715.

Jong, T., & Parry, D. L. (2003). Removal of sulfate and heavy metals by sulfate-reducing bacteria in short term bench scale upflow anaerobic packed bed reactor runs. *Water Res., 37*, 3379-3389.

Karnachuk, O. V., Kurochkina, S. Y., & Tuovinen, O. H. (2002). Growth of sulfate-reducing bacteria with solid-phase electron acceptors. *Appl. Microbiol. Biotechnol., 58*, 482-486.

Kijjanapanich, P., Pakdeerattanamint, K., Lens, P. N. L., & Annachhatre, A. P. (2012). Organic substrates as electron donors in permeable reactive barriers for removal of heavy metals from acid mine drainage. *Environ. Technol., 33*(23), 2635-2644.

Lens, P. N. L., & Kuenen, J. G. (2001). The biological sulfur cycle: novel opportunities for environmental biotechnology. *Water Sci. Technol., 44*(8), 57-66.

Lens, P. N. L., Vallero, M., Esposito, G., & Zandvoort, M. (2002). Perspectives of sulfate reducing bioreactor in environmental biotechnology. *Rev. Environ. Sci. Biotechnol., 1*, 311-325.

Liamleam, W. (2007). *Zinc removal from industrial discharge using thermophilic biological sulfate reduction with molasses as electron donor.* Asian Institute of Technology, Thailand.

Liamleam, W., & Annachhatre, A. P. (2007). Electron donors for biological sulfate reduction. *Biotechnol. Adv., 25*(5), 452-463.

Montero, A., Tojo, Y., Matsuto, T., Yamada, M., Asakura, H., & Ono, Y. (2010). Gypsum and organic matter distribution in a mixed construction and demolition waste sorting process and their possible removal from outputs. *J. Hazard. Mater., 175*, 747-753.

O'Flaherty, V., & Colleran, E. (1999). Effect of sulphate addition on volatile fatty acid and ethanol degradation in an anaerobic hybrid reactor I: process disturbance and remediation. *Bioresour. Technol., 68*(2), 101-107.

Ruwisch, R. C. (1985). A qiuck method for the determination of dissolved and precipitated sulfides in cultures of sulfate-reducing bacteria. *J. Microbiol. Methods, 4*, 33-36.

Sawyer, C. N., McCarty, P. L., & Parkin, G. F. (2003). *Chemistry for environmental engineering and science* (5th ed.): Mc Graw-Hill International

Shternina, E. B. (1960). Solubility of gypsum in aqueous solutions of salts. *Int. Geol. Rev., 1*, 605-616.

Speece, R. E. (1983). Anaerobic biotechnology for industrial wastewater treatment. *Environ. Sci. Technol., 17*, 416A-427A.

Stevens, W. (2013). Personal Communication www.smink-groep.nl.

Townsend, T., Tolaymat, T., Leo, K., & Jambeck, J. (2004). Heavy metals in recovered fines from construction and demolition debris recycling facilities in Florida. *Sci. Total Environ., 332*, 1-11.

Turley, W. (1998). What's happening in gypsum recycling. *C&D Debris Recycling, 5*(1), 8-12.

U.S.EPA. (1998). *Characterization of building-related construction and demolition debris in the United States*. Washington D.C.

van Den Ende, J. (1991). Supersaturation of soil solutions with respect to gypsum. *Plant and Soil, 133*, 65-74.

van Driessche, A. E. S., Benning, L. G., Rodriguez-Blanco, J. D., Ossorio, M., Bots, P., & Gatcia-Ruiz, J. M. (2012). The role and implication of bassanite as a stable precursor phase to gypsum precipitation. *Sci., 336*(6077), 69-72.

Verheye, W. H., & Boyadgiev, T. G. (1997). Evaluating the land use potential of gypsiferous soils from field pedogenic characteristics. *Soil Use Manage., 13*, 97-103.

Villa-Gomez, D. K., Cassidy, J., Keesman, K., Sampaio, R., & Lens, P. N. L. (2013). Tuning strategies to control the sulfide concentration using a pS electrode in sulfate reducing bioreactor. *Water Res., Submitted*.

Vincke, E., Boon, N., & Verstraete, W. (2001). Analysis of the microbial communities on corroded concrete sewer pipes - a case study. *Appl. Microbiol. Biotechnol., 57*, 776-785.

Whiffin, V. S., van Paassen, L. A., & Harkes, M. P. (2007). Microbial carbonate precipitation as a soil improvement technique. *Geomicrobiol. J., 24*, 417-423.

CHAPTER 6

Biological Sulfate Removal from Construction and Demolition Debris Leachate: Effect of Bioreactor Configuration

This chapter has been published as:
Kijjanapanich, P., Do, A. T., Annachhatre, A. P., Esposito, G., Yeh, D. H., & Lens, P. N. L. (2013). Biological sulfate removal from construction and demolition debris leachate: Effect of bioreactor configuration. *J. Hazard. Mater. [G16 Special Issue], In Press*. DOI: 10.1016/j.jhazmat.2013.10.015

Chapter 6

Due to the contamination of construction and demolition debris (CDD) by gypsum drywall, especially, its sand fraction (CDD sand, CDDS), the sulfate content in CDDS exceeds the posed limit of the maximum amount of sulfate present in building sand (1.73 g sulfate per kg of sand for The Netherlands). Therefore, the CDDS cannot be reused for construction. The CDDS has to be washed in order to remove most of the impurities and to obtain the right sulfate content, thus generating a leachate, containing high sulfate and calcium concentrations. This study aimed at developing a biological sulfate reduction system for CDDS leachate treatment and compared three different reactor configurations for the sulfate reduction step: the Upflow Anaerobic Sludge Blanket (UASB) reactor, Inverse Fluidized Bed (IFB) reactor and Gas Lift Anaerobic Membrane Bioreactor (GL-AnMBR). This investigation demonstrated that all three systems can be applied for the treatment of CDDS leachate. The highest sulfate removal efficiency of 75-85% was achieved at a hydraulic retention time (HRT) of 15.5 h. A high calcium concentration up to 1000 mg L^{-1} did not give any adverse effect on the sulfate removal efficiency of the IFB and GL-AnMBR systems.

6.1. Introduction

CDD originates from building, demolition and renovation of buildings. Due to insufficient source separation, CDD becomes a mixed material which is difficult to recycle (Montero et al., 2010). The composition of CDD is affected by numerous factors, including the raw materials used, architectural techniques, local construction and demolition practices (Dorsthorst & Kowalczyk, 2002). The main ingredients present in the CDD are soil, ballast, concrete, asphalt, bricks, tiles, masonry, wood, metals, paper, plastics and gypsum drywall (Dorsthorst & Kowalczyk, 2002; Thomson, 2004; U.S.EPA, 1998). Moreover, toxic wastes, like asbestos and heavy metals, are not always separated from the rest of the CDD. Although their quantity is relatively small, their presence can significantly affect the recycled materials or can contaminate landfills (Dorsthorst & Kowalczyk, 2002).

According to several characterization studies of CDD in the US, gypsum drywall accounts for 21-27% of the mass of debris generated during the construction and renovation of residential structures (U.S.EPA, 1998). On an average, 0.9 metric tons of waste gypsum is generated from the construction of a typical single family home or 4.9 kg m^{-2} of the structure (Turley, 1998). Nearly 40% of the total mass of CDD is CDD sand (CDDS), which consists mainly of sand (de Vries, 2006), due to its weight and extensive usage in modern building techniques. Moreover, most of the gypsum is concentrated in the sand fraction (52.4% of total gypsum) (Montero et al., 2010; Townsend et al., 2004), whereas the organic matter is distributed mainly in the large-sized fractions of CDD (Montero et al., 2010).

Reuse options have been proposed for CDDS, including soil amendment, alternative daily landfill cover, and fill material in roads, embankment and construction projects. The presence of gypsum drywall in CDDS may provide some benefits as a soil conditioner or nutrient source. However, for applications where the material is placed in direct contact with the environment, a concern has been raised by regulators regarding the chemical characteristics of the material and the potential risk to human health and the environment (Jang & Townsend, 2001). In the EU, about 75% of the 'core' CDD is nowadays landfilled, while only 25% is reused (Montero et al., 2010). In addition, the EU has recently introduced targets for CDD, according to which a 70% recycling target (the EU Waste Framework Directive 2008/98/EC) has to be achieved by 2020 (Tojo & Fischer, 2011). Recycling percentages in the EU (Table 6.1) vary from 0.7% (Cyprus) to more than 80% (Germany, Estonia, Denmark and the Netherlands) (Monier et al., 2011; Tojo & Fischer, 2011).

Table 6.1. Recycling percentages of construction and demolition debris (CDD) of the EU countries in 2004-2006 (Dorsthorst & Kowalczyk, 2002; Monier et al., 2011; Tojo & Fischer, 2011)

The EU countries	Total recycling of CDD	
	Tons per capita	Percentages
Netherlands	1.55	98.1
Denmark	1.07	94.9
Estonia	1.64	91.9
Germany	1.93	86.3
Ireland	3.14	79.5
Belgium	0.75	67.5
United Kingdom	1.22	64.8
France	3.42	62.3
Norway	0.16	61.0
Lithuania	0.11	59.7
Austria	0.48	59.5
Latvia	0.02	45.8
Poland	0.13	28.3
Finland	0.41	26.3
Czech Republic	0.27	23.0
Hungary	0.08	15.5
Spain	0.12	13.6
Cyprus	0.01	0.7

The Netherlands has drawn up a national "Building site waste" plan comprising measures aimed at banning the landfilling of recoverable waste (Dorsthorst & Kowalczyk, 2002). Nowadays in the Netherlands about 98% of the CDD is recovered and reused (Monier et al., 2011). Since January 2001, it is forbidden to dump reusable and combustible CDD on a landfill (Dorsthorst & Kowalczyk, 2002). The Dutch government has set the limits to the maximum amount of polluting compounds present in building material. For reusable sand, the emission limit is set to 1.73 g sulfate per kg of sand (de Vries, 2006; Stevens, 2013). However, most of the CDDS still remains highly polluted, and the sulfate content often exceeds the prescribed limit (de Vries, 2006).

Processes for sulfate removal from CDDS have been developed. CDDS is washed to remove most of the impurities, to obtain the right physical characteristics (de Vries, 2006; Kijjanapanich et al., 2013a; Kijjanapanich et al., 2013b) and also to leach out the gypsum from the material. A novel approach for the removal of sulfate based on the biological treatment of sulfate containing wastewater (Annachhatre & Suktrakoolvait, 2001; Benner et al., 1999; Costa et al., 2007; Waybrant et al., 1998), has been proposed and also be applied for the treatment of CDDS leachate (Kijjanapanich et al., 2013a). This approach uses the bacterial sulfate reduction process as it occurs in nature for the removal of sulfate, often coupled to heavy metal removal (Jong & Parry, 2003;

Kijjanapanich et al., 2012; Liamleam, 2007). Many types of bioreactors have been used for the sulfate reduction step, including the UASB Reactor, Fluidized Bed Reactor (FBR), IFB Reactor, Continuous Stirred Tank Reactor (CSTR) and Anaerobic Membrane Bioreactor (AnMBR) (Annachhatre & Suktrakoolvait, 2001; Kijjanapanich et al., 2013a; Nevatalo et al., 2010; Sahinkaya et al., 2011; Vallero et al., 2005; Villa-Gomez et al., 2011). The selection of a reactor configuration is often determined by the type of wastewater to be treated, possible advantages and disadvantages of the reactors, its operational cost and reliability (Hatzikioseyian & Remoundaki; Ram et al., 1993).

Research on bioremediation of CDDS leachate, especially using sulfate reducing bacteria (SRB) is rare. Therefore, this research aimed to study a biological sulfate reduction system to reduce the sulfate content of CDDS leachate using three different types of bioreactors. The effect of the calcium concentration contained in the CDDS leachate on the sulfate removal efficiency was also investigated. First, a UASB reactor was selected as it is the most widely applied reactor configuration for anaerobic wastewater treatment throughout the world (Lettinga, 1996). The IFB is a promising reactor configuration for the combined biological sulfate reduction and metal precipitate separation in a single reactor unit (Villa-Gomez et al., 2011). The last reactor configuration studied was a GL-AnMBR which is suitable for slow growing microorganisms, has a smaller reactor footprint and produces excellent effluent quality (Lee & Kim, 2009). Moreover, the gas lift system of this AnMBR may alleviate the sulfide toxicity on SRB.

6.2. Material and Methods
6.2.1 Construction and demolition debris (CDD)

CDD samples were collected from Smink Afvalverwerking B.V. (Amersfoort, The Netherlands). Samples were air-dried and sieved at 2 mm. Pieces of wood, concrete, rock, paper, plastic and foam were removed, thus retaining only the sand fraction (CDDS).

6.2.2 Sulfate reducing bacteria (SRB) inoculums

Mixed anaerobic granular sludge provided by Biothane Systems International (Delft, The Netherlands) was used as source for SRB in UASB. The seed sludge had a TSS and VSS content of 54.6 g L^{-1} and 39.8 g L^{-1}, respectively, corresponding to a VSS/TSS ratio of 0.73. Anaerobic sludge from a digester treating activated sludge from a domestic wastewater treatment plant (De Nieuwe Waterweg in Hoek van Holland, The Netherlands) was used as source for SRB in IFB and GL-AnMBR.

6.2.3 Construction and demolition debris sand (CDDS) leachate

The leaching columns were made of polyvinyl chloride (PVC) which had a working volume of 2 L. One kg of CDDS was filled in leaching column. Demineralized water was fed at the bottom of the column at a flow rate of 252 mL h^{-1} (0.1 m h^{-1}) using a peristaltic pump. The leachate was withdrawn from the top of the column. The experiments were conducted at room temperature (23 ± 4°C). The CDDS leachate was then diluted to a sulfate concentration of 600 mg L^{-1} before feeding to all bioreactors. The characteristics of this CDDS leachate (Solid:Liquid ratio = 1:10) was described in the study of Kijjanapanich et al. (2013a).

6.2.4 Bioreactor configurations

6.2.4.1 Upflow anaerobic sludge blanket (UASB) reactor

The UASB reactor was made of polymethylmethacrylate (PMMA) and had a working volume of 3.9 L (Figures 6.1a and 6.1d). The reactor contained 50% by volume of anaerobic granular sludge. The CDDS leachate was fed at the bottom of the UASB reactor at a flow rate of 252 mL h^{-1} using a peristaltic pump. The effluent was withdrawn from the top of the column.

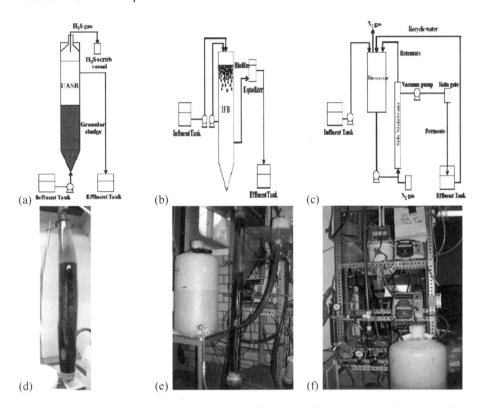

Figure 6.1. Schematic diagram of the bioreactors (Above: reactor schematic and Below: lab-scale bioreactors): (a, d) Upflow Anaerobic Sludge Blanket (UASB) reactor, (b, e) Inverse Fluidized Bed (IFB) Reactor and (c, f) Gas Lift Anaerobic Membrane Bioreactor (GL-AnMBR).

6.2.4.2 Inverse fluidized bed (IFB) reactor

The IFB reactor was made of PMMA and it had a working volume of 2.5 L (Figures 6.1b and 6.1e). The carrier material consisted of 600 mL low-density polyethylene beads (PurellPe 1810 E, BasellPolylifins, The Netherlands) of 3 mm diameter. The reactor start-up was accomplished as per the procedure developed in the study of Villa-Gomez et al. (2011). The expansion of the bed was maintained at 30% of the reactor volume by means of the recirculation flow using a magnetic drive pump. The effluent was withdrawn from the equalizer tank, which was connected at the upper part of the reactor in order to maintain a constant liquid level in the IFB (Villa-Gomez et al., 2011).

6.2.4.3 Gas lift anaerobic membrane bioreactor (GL-AnMBR)

A Norit-X flow (F4785) tubular PVDF hollow fiber membrane with a pore size of 0.03 μm (Norit Membrane Technology, Enschede, The Netherlands) was used (Prieto, 2011). The schematic of the GL-AnMBR is shown in Figures 6.1c and 6.1f. The bioreactor was made of PMMA with a working volume of 1.8 L. The TSS and VSS concentrations in the reactor were 6000-7000 mg L^{-1} and 5000-6000 mg L^{-1}, respectively. The membrane feed pump set at 40 L h^{-1} transported the sludge from the bottom of the reactor to the membrane section. Nitrogen gas was used to lift up the sludge (with the enhancement via the membrane pump) at a gas flow rate of 0.15 L min^{-1}.

A permeate flow rate of 2 mL min^{-1} was maintained using a peristaltic pump in order to withdraw the water through the membrane (Prieto, 2011). The retentate water then was recycled back to the reactor. The permeate water flow rate was determined by a rain gauge. After passing through the rain gauge, the permeate was pumped out at the same rate as the reactor feed flow rate. The membrane was cleaned by tap water every day for 15 min and backwashed for 10-15 min (Prieto, 2011; Prieto et al., 2013).

6.2.5 Bioreactor experiments

All three bioreactor configurations (UASB, IFB and GL-AnMBR) were continuously fed with the CDDS leachate and operated for 45 d in case of UASB and 60 d for the other two reactors. Ethanol was supplied as electron donor at an organic loading rate (OLR) of 1.75-2.17 g COD L^{-1} reactor d^{-1} in all the three reactors. The CDDS leachate had a sulfate concentration of 600 mg L^{-1}, corresponding to a sulfate loading rate of 0.93 g sulfate L^{-1} reactor d^{-1}. No pH adjustment was carried out in any of the bioreactors. Table 6.2 shows the operation conditions of the three bioreactors.

From day 47 until the end of the experiment (day 60), calcium chloride was added to the CDDS leachate to achieve a calcium concentration of 1000 mg L^{-1} in order to study the effect of the calcium concentration on the sulfate removal efficiency in the IFB and GL-AnMBR. Clogging of the sludge bed due to the precipitation of calcium carbonate onto the surface of the UASB granular sludge occurred (data not shown). Therefore, the effect of the calcium concentration was not further tested in case of the UASB reactor.

Table 6.2. Operational conditions applied to the bioreactors used in this study

Parameters	Systems		
	UASB	IFB	GL-AnMBR
Working volume (L)	3.9	2.5	1.8
Temperature (°C)	23 ± 4	23 ± 4	23 ± 4
HRT (h)	15.5	15.5	15.5
Water flow rate (mL h^{-1})	252	161	116
Upflow velocity (m h^{-1})	0.1	-	-
OLR (g COD L^{-1}reactor d^{-1})	1.75	1.75	2.17
SLR (g sulfate L^{-1}reactor d^{-1})	0.93	0.93	0.93

HRT: hydraulic retention time, OLR: organic loading rate, SLR: sulfate loading rate

6.2.6 Analytical methods

The pH was measured as overall acidity indicator using a 691 Metrohm pH meter and a SenTix 21 WTW pH electrode, while the oxidation-reduction potential (ORP) was measured as redox condition indicator of the system using a 340i WTW pH meter and a QR481X QIS ORP electrode. Sulfate was measured using an ICS-1000 Dionex Ion Chromatography (IC) (Eaton et al., 2005). Sulfide was measured by the method of Ralf Cord-Ruwisch (Ruwisch, 1985) using a Perkin Elmer Lambda 20 UV visible spectrophotometer. Calcium was measured using an AAnalyst 200 Perkin Elmer Atomic Absorption Spectrometer (AAS)-Flame (Eaton et al., 2005). Ethanol and acetate were measured using a Varian 430 Gas Chromatograph (GC) (Eaton et al., 2005). The dissolved organic carbon (DOC) was monitored as an indicator of dissolved carbon available for bacterial activity. The DOC was measured using the high temperature combustion method by Shimadzu TOC-V CPN analyzer (Eaton et al., 2005).

6.3. Results
6.3.1 Upflow anaerobic sludge blanket (UASB) reactor

The UASB reactor influent and effluent pH values were maintained around the neutral range (pH 6.0-8.6) without pH adjustment (Figure 6.2a). The UASB reactor effluent pH (average pH 8.4) was higher than the influent pH (average pH 6.8). The ORP of the UASB reactor effluent was between -211 and -405 mV throughout the experiment. There was an increase in the effluent sulfide concentration from 10 to 200 mgS L^{-1} (Figure 6.2b) as the experiment progressed, in congruence with an increase of the sulfate removal efficiency (Figure 6.2c).

The sulfate removal efficiency of the UASB reactor was around 30-50% during the first 30 d of the experiment beyond which the sulfate concentration of the UASB effluent decreased rapidly, reaching a sulfate removal efficiency up to 82% within 10 d (Figure 6.2c). The calcium concentrations in the UASB reactor influent and effluent remained almost equal (Figure 6.2d). No ethanol was detected in the UASB effluent, while the acetate concentration in the effluent was around 200 mg L^{-1} throughout the experiment (Figure 6.2e). Consequently, the average DOC removal efficiency achieved was only about 50% (Figure 6.2f).

6.3.2 Inverse fluidized bed (IFB) reactor

The IFB reactor influent and effluent pH values were maintained around the neutral range (pH 6.0-7.1) without pH adjustment (Figure 6.3a). No significant differences between the influent and effluent pH were observed. The ORP of the IFB reactor effluent remained between -308 and -379 mV throughout the experiment. Effluent sulfide concentrations varied between 40 and 260 mgS L^{-1} (Figure 6.3b).

The sulfate removal efficiency of the IFB reactor improved rapidly during the first 10 d of the experiment and increased up to 70% on day 14. Then, the sulfate removal efficiency stabilized around 75% until the end of the experiment (Figure 6.3c). The IFB reactor influent and effluent calcium concentrations were almost equal, even when a high calcium concentration (1000 mg L^{-1}) was supplied to the reactor (Figure 6.3d).

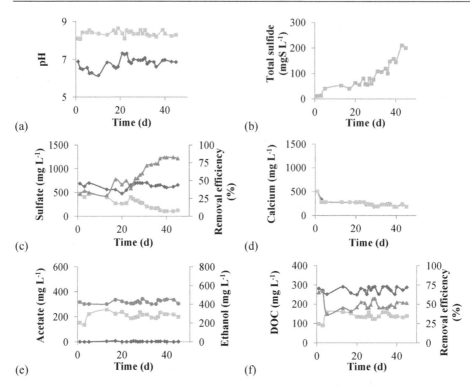

Figure 6.2. Performance of a sulfate reducing UASB reactor as a function of operation time: (a) pH, (b) total sulfide, (c) sulfate, (d) calcium (e) ethanol and acetate and (f) DOC. (♦) influent, (■) effluent, (▲) removal efficiency and (●) influent ethanol.

There was no ethanol detected in the IFB reactor effluent, while the acetate concentration in the effluent was around 300 mg L^{-1} after 15 d of operation until the end of the experiment (Figure 6.3e). The DOC removal efficiency was very high (up to 95%) at the first 15 d. Then, it decreased and the average DOC removal efficiency achieved was only 35% (Figure 6.3f).

6.3.3 Gas lift anaerobic membrane bioreactor (GL-AnMBR)

During the first 45 d, the trans-membrane pressure (TMP) across the membrane and the flux were in the range of 1.2-1.4 bar and 7.5-8.6 L m^{-2} h^{-1} (LMH), respectively (Figure 6.4g). However, after 45 d, the TMP increased to 1.9 bars, while the flux decreased to 4.5 LMH.

The GL-AnMBR influent and effluent pH values were maintained in the neutral range (pH 6.0-8.5) without pH adjustment (Figure 6.4a). The GL-AnMBR effluent pH (average pH 7.8) was higher than the influent pH (average pH 6.4). The ORP of the GL-AnMBR effluent stabilized between -300 and -350 mV throughout the experiment. The sulfide concentration in the effluent was much lower (average 10 mgS L^{-1}) (Figure 6.4b), compared to the other two reactor configurations investigated.

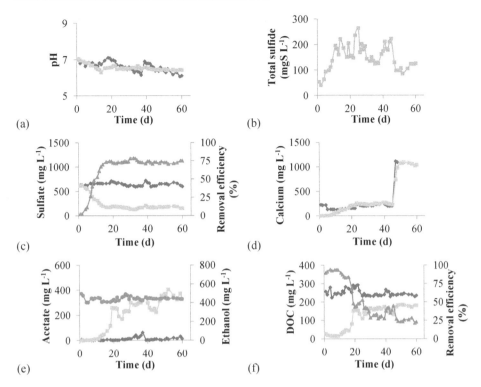

Figure 6.3. Performance of a sulfate reducing IFB reactor as a function of operation time: (a) pH, (b) total sulfide, (c) sulfate, (d) calcium (e) ethanol and acetate and (f) DOC. (♦) influent, (■) effluent, (▲) removal efficiency and (●) influent ethanol.

The sulfate removal efficiency of the GL-AnMBR stabilized around 60-80% throughout the experiment (Figure 6.4c). The GL-AnMBR influent and effluent calcium concentrations were almost the same, even when the highest calcium concentration (1000 mg L^{-1}) was supplied to the reactor (Figure 6.4d).

There was no ethanol detected in the GL-AnMBR effluent, while the acetate concentration in the effluent was around 230 mg L^{-1} (Figure 6.4e). Consequently, the average DOC removal efficiency achieved was 65% (Figure 6.4f).

6.4. Discussion
6.4.1 Sulfate removal efficiency

This study demonstrated that all three biological sulfate reduction reactor configurations (UASB, IFB and GL-AnMBR) are effective for the treatment of CDDS leachate, all of them achieving a sulfate removal efficiency of 75-85%. The sulfate removal efficiency of the IFB (75%) was slightly lower compared to the UASB and GL-AnMBR systems (80%) at the stationary phase. However, the sulfate removal efficiency was low at the beginning of the experiment, due to the acclimatization period. The GL-AnMBR system needed the shortest acclimatization time (15 d), followed by the IFB (20 d) and UASB systems (35 d). This is because the GL-AnMBR system had been operated with the CDDS leachate before with a longer HRT, while the other two systems had never

operated with this CDDS leachate before. Moreover, different inoculum seed sludge
was used for each reactor (granular sludge for UASB and biofilm for IFB)

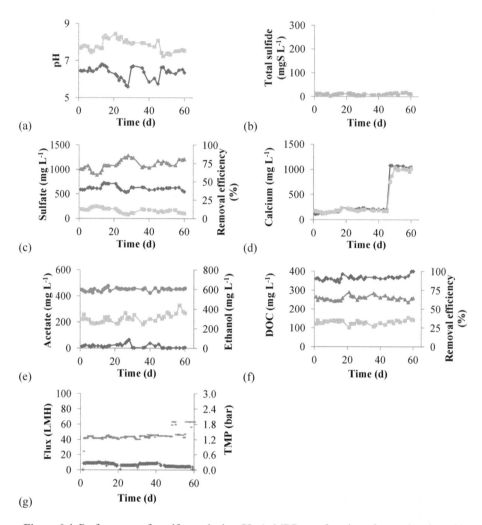

Figure 6.4. Performance of a sulfate reducing GL-AnMBR as a function of operation time: (a)
pH, (b) total sulfide, (c) sulfate, (d) calcium (e) ethanol and acetate, (f) DOC and (g) flux and
TMP. (♦) influent, (■) effluent, (▲) removal efficiency, (●) influent ethanol, (+) flux and (-)
TMP.

The composition of CDDS is affected by numerous factors, depending on the source of
the CDD (Townsend et al., 2004). According to the studies of de Vries (2006) and
Azabou (2005), SRB growth and activities can be inhibited due to impurities such as
heavy metals present in the gypsum waste. Therefore, further research about the effect
of the impurities from CDDS on the biological sulfate reduction is recommended to
assess if an appropriate pre-treatment of CDDS leachates is required.

6.4.2 pH and dissolved organic carbon (DOC) removal efficiency

The effluent pH of the UASB and GL-AnMBR systems, which was higher than the influent pH (Figures 6.2a and 6.4a), indicates the generation of alkalinity during the sulfate reduction process (Sawyer et al., 2003). In contrast, the effluent pH of IFB system was almost similar to the influent pH (6.0-7.1). This might be due to the accumulation of acetate in the system (Figure 6.3e), which consumed the alkalinity generated by the SRB.

For biological sulfate reduction, a carbon source and an electron donor are the primary requirements for SRB. Ethanol was used because it is cheap and easy to use. A sulfate removal efficiency as high as 80% has been achieved at high sulfate loading rates (up to 10-12 gS L^{-1} d^{-1}) with ethanol as electron donor (de Smul et al., 1997). The use of organic waste as a cheap electron donor for the SRB is also an interesting option, as many the companies have such waste streams and coupling bio-waste treatment with CDD clean-up is an interesting example of how enhanced resource recovery can contribute to sustainable development.

When using ethanol as electron donor, especially in high rate systems, high concentrations of acetate (up to 1 g L^{-1}) are generated which results in an increase in effluent COD (Cao et al., 2012; Kijjanapanich et al., 2013a; Liamleam & Annachhatre, 2007). Acetate production during biological sulfate reduction is nevertheless a major drawback of sulfate reducing reactors. Most SRBs cannot completely oxidize acetate even with excess sulfate and the enrichment of acetate oxidizing SRB requires a longer time because of their slow growth rate (Lens et al., 2002). In this study, there was acetate accumulation in the IFB system (Figure 6.3e), resulting in the effluent pH lower than those in the other two systems (Figure 6.3a). The acetate concentration in the effluent was constant at around 200 mg L^{-1} in case of the UASB and GL-AnMBR systems (Figures 6.2e and 6.4e).

GL-AnMBR gave the highest DOC removal efficiency in this study (Figure 6.4f), even though a higher OLR was applied. This may be due to the membrane retention of dissolved organic compounds inside the reactor, resulting in a longer retention time (biodegradation time) and thus lower DOC in the effluent.

6.4.3 Effect of calcium on the sulfate removal efficiency and bioreactor operation

The UASB influent and effluent calcium concentrations remained almost equal (Figure 6.2d), but precipitation of calcite on the granular sludge was observed (data not shown), resulting in severe agglomeration of the sludge in the UASB reactor. Treatment of wastewater with high calcium concentrations (780-1560 mg L^{-1}) using a UASB reactor results in the rapid formation of dense granules with a high ash content which easily agglomerates, leading to serious cementation of the sludge bed (van Langerak et al., 2000). Therefore, the UASB system was not further tested with higher influent calcium concentrations (1000 mg L^{-1}).

The calcium concentration up to 1000 mg L^{-1} did not affect the sulfate reduction process in both IFB and GL-AnMBR systems and the sulfate removal efficiency remained constant until the end of the experiment (Figures 6.3c and 6.4c). However, calcium precipitation as calcium carbonate ($CaCO_3$) might occur, which can accumulate and can eventually affect on the sulfate reduction. Therefore, a study of the effect of higher calcium concentrations on long term reactor operation is required. A calcium recovery step may be required to prevent accumulation of $CaCO_3$ precipitates in the piping or polyethylene beads, e.g. by either chemical precipitation (Benefield & M., 1999) or microbial carbonate precipitation (MCP) using ureolytic bacteria (Al-Thawadi & Cord-Ruwisch, 2012; Al-Thawadi, 2011; Hammes et al., 2003; Whiffin et al., 2007).

In the GL-AnMBR system, the TMP increased to 1.9 bars, and the flux decreased to 4.5 LMH at the end of the experiment. This might be explained from the fact that after operation for a certain time, the membrane pores get compact and blocked by small chemical ($CaCO_3$) or biological particles, which results in an increase of the TMP as well as a decrease of the flux. At the end of the experiment, the membrane was washed with 2 L of hydrochloric acid (HCl) solution (pH 4) and the leachate was then analyzed for its calcium concentration. The calcium concentration from this washing process was only 188 mg, corresponding to 0.29% of the total calcium loaded to the system. Thus, little scaling of the membrane occurred. Since day 30, a decrease of the TSS concentration in the reactor was also observed (data not shown). However, the TSS concentration increased again after backwashing process (data not shown). Therefore, the increase of TMP and decrease of the flux is likely due to biological fouling.

6.4.4 Sulfide production

There was a significant fluctuation in the effluent sulfide concentration of the IFB reactor (Figure 6.3b), as compared to the UASB reactor effluent sulfide concentration. The sulfide concentration is pH dependent (Sawyer et al., 2003). Under acidic conditions sulfide combines with protons to produce H_2S, which may be released as gaseous product (Jong & Parry, 2003). In this study, the pH of the IFB reactor effluent was around 6.5 so around 50% of the sulfide was in the H_2S form, while the pH of the UASB reactor effluent was up to 8.6. This implies that the percentage of H_2S in UASB effluent was lower as compare to the H_2S percentage in the IFB reactor effluent, resulting in less fluctuation of the sulfide concentration. Moreover, the room temperature (23 ± 4°C) also slightly oscillated during the experiment, thus inducing fluctuations in the sulfate reduction rates and subsequent effluent sulfide concentrations.

The effluent of the GL-AnMBR showed the lowest sulfide concentration (Figure 6.4b), as compared to UASB and IFB reactor effluents. Nitrogen gas was used to lift up the sludge in the GL-AnMBR system, thus the sulfide could be easily stripped from the reactor liquor to the gas phase as H_2S. This resulted in lower GL-AnMBR effluent sulfide concentrations in ranging from 3 to 18 $mgS\ L^{-1}$. This was expected to reduce the sulfide toxicity on SRB. However, the sulfate removal efficiency did not differ much from the other systems. This is because the sulfide production still did not reach toxic levels of SRB. A higher sulfate removal efficiency of the GL-AnMBR compared to the other two systems should be found when operating at higher sulfate loading rates, resulting in increased sulfide production.

The sulfide produced during this biological process can also be used for heavy metal removal from the leachate itself as well as from other wastewaters (Jong & Parry, 2003; Kijjanapanich et al., 2012; Liamleam, 2007). Alternatively, it can be used for recovery of elemental sulfur (S^0) (Dutta et al., 2008) or sulfuric acid (H_2SO_4) (Laursen & Karavanov, 2006).

6.5. Conclusions

This investigation demonstrated that three bioreactor configurations, an UASB, IFB and GL-AnMBR, can be used for the treatment of CDDS leachate (up to 75-85% sulfate removal efficiency), where SRB use the sulfate in CDDS leachate as a source of sulfate. A high calcium concentration had an adverse impact on the UASB granular sludge, as $CaCO_3$ precipitation occurred on the UASB granule surface. On the other hand, a calcium concentration up to 1000 mg L^{-1} did not have any adverse effect on the sulfate removal efficiency of the IFB and GL-AnMBR systems. The effluent of these bioprocesses still had high sulfide (except GL-AnMBR) and calcium concentration, which have to be removed prior to reuse as water in the leaching process or discharge it to the environment.

6.6 References

Al-Thawadi, S., & Cord-Ruwisch, R. (2012). Calcium carbonate crystals formation by ureolytic bacteria isolated from Australian soil and sludge. *J. Adv. Sci. Eng. Res., 2*, 12-26.

Al-Thawadi, S. M. (2011). Ureolytic bacteria and calcium carbonate formation as a mechanism of strength enhancement of sand. *J. Adv. Sci. Eng. Res., 1*, 98-114.

Annachhatre, A. P., & Suktrakoolvait, S. (2001). Biological sulfate reduction using molasses as a carbon source. *Water Environ. Res., 73*, 118-126.

Azabou, S., Mechichi, T., & Sayadi, S. (2005). Sulfate reduction from phosphogypsum using a mixed culture of sulfate-reducing bacteria. *Int. Biodeter. Biodegr., 56*(4), 236-242.

Benefield, L. D., & M., M. J. (1999). Chemical precipitation. In American Water Works Association (Ed.), *Water Quality and Treatment: A Handbook of Community Water Supplies 5th ed.* (Fifth ed.): McGraw-Hill, Inc.

Benner, S. G., Blowes, D. W., Gould, W. D., Herbert, R. B., & Ptacek, C. J. (1999). Geochemistry of a permeable reactive barrier for metals and acid mine drainage. *Environ. Sci. Technol., 33*(16), 2793-2799.

Cao, J., Zhang, G., Mao, Z., Li, Y., Fang, Z., & Yang, C. (2012). Influence of electron donors on the growth and activity of sulfate-reducing bacteria. *Int. J. Miner. Process., 106-109*, 58-64.

Costa, M. C., Martins, M., Jesus, C., & Duarte, J. C. (2007). Treatment of acid mine drainage by sulfate-reducing bacteria using low cost matrices. *Water Air Soil Pollut., 189*(1-4), 149-162.

de Smul, A., Dries, J., Goethals, L., Grootaerd, H., & Verstraete, W. (1997). High rates of microbial sulphate redction in a mesophilic ethanol-fed expanded-granular-sludge-blanket reactors. *Appl. Microbiol. Biotechnol., 48*, 297-303.

de Vries, E. (2006). *Biological sulfate removal from construction and demolition debris sand.* Wageningen University.

Dorsthorst, B. J. H., & Kowalczyk, T. (2002). *Design for Recycling.* Paper presented at the Design for Deconstruction and Materials Reuse, The CIB Task Group 39 – Deconstruction Meeting, Karlsruhe, Germany.

Dutta, P. K., Rabaey, K., Yuan, Z., & Keller, J. (2008). Spontaneous electrochemical removal of aqueous sulfide. *Water Res., 42*, 4965-4975.

Eaton, A. D., APHA, AWWA, & WEF. (2005). *Standard methods for the examination of water and wastewater* (21st ed.). Washington D.C.

Hammes, F., Seka, A., de Knijf, S., & Verstraete, W. (2003). A novel approach to calcium removal from calcium-rich industrial wastewater. *Water Res., 37*(3), 699-704.

Hatzikioseyian, A., & Remoundaki, E. Bioreactor for metal bearing wastewater treatment. Retrieved 7 June, 2012, from http://www.metal.ntua.gr/~pkousi/e-learning/bioreactors/page_23.htm

Jang, Y. C., & Townsend, T. (2001). Sulfate leaching from recovered construction and demolition debris fines. *Adv. Environ. Res., 5*, 203-217.

Jong, T., & Parry, D. L. (2003). Removal of sulfate and heavy metals by sulfate-reducing bacteria in short term bench scale upflow anaerobic packed bed reactor runs. *Water Res., 37*, 3379-3389.

Kijjanapanich, P., Annachhatre, A. P., Esposito, G., van Hullebusch, E. D., & Lens, P. N. L. (2013a). Biological sulfate removal from gypsum contaminated construction and demolition debris. *J. Environ. Manage., 131*, 82-91.

Kijjanapanich, P., Annachhatre, A. P., & Lens, P. N. L. (2013b). Biological sulfate reduction for treatment of gypsum contaminated soils, sediments and solid wastes. *Crit. Rev. Environ. Sci. Technol., In Press.*

Kijjanapanich, P., Pakdeerattanamint, K., Lens, P. N. L., & Annachhatre, A. P. (2012). Organic substrates as electron donors in permeable reactive barriers for removal of heavy metals from acid mine drainage. *Environ. Technol., 33*(23), 2635-2644.

Laursen, J. K., & Karavanov, A. N. (2006). Processes for sulfur recovery, regeneration of spent acid, and reduction of NO_x emissions. *Chem. Pet. Eng., 42*(5-6), 229-234.

Lee, M., & Kim, J. (2009). Membrane autopsy to investigate $CaCO_3$ scale formation in pilot-scale, submerged membrane bioreactor treating calcium-rich wastewater. *J. Chem. Technol. Biotechnol., 84*(9), 1397-1404.

Lens, P. N. L., Vallero, M., Esposito, G., & Zandvoort, M. (2002). Perspectives of sulfate reducing bioreactor in environmental biotechnology. *Rev. Environ. Sci. Biotechnol., 1*, 311-325.

Lettinga, G. (1996). Sustainable integrated biological wastewater treatment. *Water Sci. Technol., 33*, 85-98.

Liamleam, W. (2007). *Zinc removal from industrial discharge using thermophilic biological sulfate reduction with molasses as electron donor.* Asian Institute of Technology, Thailand.

Liamleam, W., & Annachhatre, A. P. (2007). Electron donors for biological sulfate reduction. *Biotechnol. Adv., 25*(5), 452-463.

Monier, V., Hestin, M., Trarieux, M., Mimid, S., Domröse, L., van Acoleyen, M., et al. (2011). *Study on the management of construction and demolition waste in the EU.*

Montero, A., Tojo, Y., Matsuto, T., Yamada, M., Asakura, H., & Ono, Y. (2010). Gypsum and organic matter distribution in a mixed construction and demolition waste sorting process and their possible removal from outputs. *J. Hazard. Mater., 175*, 747-753.

Nevatalo, L. M., Makinen, A. E., Kaksonen, A. H., & Puhakka, J. A. (2010). Biological hydrogen sulfide production in an ethanol-lactate fed fluidized-bed bioreactor. *Bioresour. Technol., 101*, 276-284.

Prieto, A. L. (2011). *Sequential anaerobic and algal membrane bioreactor (A2MBR) system for sustainable sanitation and resource recovery from domestic wastewater.* University of South Florida, Tampa, Florida.

Prieto, A. L., Futselaar, H., Lens, P. N. L., Bair, R., & Yeh, D. H. (2013). Development and start up of a gas-lift anaerobic membrane bioreactor (Gl-AnMBR) for conversion of sewage to energy, water and nutrients. *J. Membr. Sci., 441*, 158-167.

Ram, N. M., Bass, D. H., Falotico, R., & Leahy, M. (1993). A decision framework for selecting remediation technologies at hydrocarbon-contaminated sites. *J. Soil Contam., 2*(2), 1-24.

Ruwisch, R. C. (1985). A qiuck method for the determination of dissolved and precipitated sulfides in cultures of sulfate-reducing bacteria. *J. Microbiol. Methods, 4*, 33-36.

Sahinkaya, E., Gunes, F. M., Ucar, D., & Kaksonen, A. H. (2011). Sulfidogenic fluidized bed treatment of real acid mine drainage water. *Bioresour. Technol., 102*, 683-689.

Sawyer, C. N., McCarty, P. L., & Parkin, G. F. (2003). *Chemistry for environmental engineering and science* (5th ed.): Mc Graw-Hill International

Stevens, W. (2013). Personal Communication www.smink-groep.nl.

Thomson, L. (2004). construction and demolition debris recovery program. Retrieved 06, 2011, from http://www.cccounty.us/depart/cd/recycle/debris.htm

Tojo, N., & Fischer, C. (2011). *Europe as a recycling society: European recycling policies in relation to the actual recycling achieved.*

Townsend, T., Tolaymat, T., Leo, K., & Jambeck, J. (2004). Heavy metals in recovered fines from construction and demolition debris recycling facilities in Florida. *Sci. Total Environ., 332*, 1-11.

Turley, W. (1998). What's happening in gypsum recycling. *C&D Debris Recycling, 5*(1), 8-12.

U.S.EPA. (1998). *Characterization of building-related construction and demolition debris in the United States*. Washington D.C.

Vallero, M. V. G., Lettinga, G., & Lens, P. N. L. (2005). High rate sulfate reduction in a submerged anaerobic membrane bioreactor (SAMBaR) at high salinity. *J. Membr. Sci., 253*(1-2), 217-232.

van Langerak, E. P. A., Ramaekers, H., Wiechers, J., Veeken, A. H. M., Hamelers, H. V. M., & Lettinga, G. (2000). Impact of location of $CaCO_3$ precipitation on the development of intact anaerobic sludge. *Water Res., 34*(2), 437-446.

Villa-Gomez, D., Ababneh, H., Papirio, S., Rousseau, D. P. L., & Lens, P. N. L. (2011). Effect of sulfide concentration on the location of the metal precipitates in inversed fluidized bed reactors. *J. Hazard. Mater., 192*, 200-207.

Waybrant, K. R., Blowes, D. W., & Ptacek, C. J. (1998). Selection of Reactive Mixtures for Use in Permeable Reactive Walls for Treatment of Mine Drainage. *Environ. Sci. Technol., 32*(13), 1972-1979.

Whiffin, V. S., van Paassen, L. A., & Harkes, M. P. (2007). Microbial carbonate precipitation as a soil improvement technique. *Geomicrobiol. J., 24*, 417-423.

CHAPTER 7

Chemical Sulfate Removal for Treatment of Construction and Demolition Debris Leachate

This chapter was submitted for publication as:
Kijjanapanich, P., Annachhatre, A. P., Esposito, G., & Lens, P. N. L. (2013). Chemical sulfate removal for treatment of construction and demolition debris leachate. *Environ. Technol., Submitted.*

Chapter 7

Construction and demolition debris (CDD) is a product of construction, renovation or demolition activities. It has a high gypsum content (52.4% of total gypsum), concentrated in the CDD sand fraction (CDDS). To comply with the posed limit of the maximum amount of sulfate present in building sand, excess sulfate needs to be removed. In order to enable reuse of CDDS, a novel treatment process is developed based on washing of the CDDS to remove most of the gypsum, and subsequent sulfate removal from the sulfate rich CDDS leachate. This study aims to assess chemical techniques, i.e. precipitation and adsorption, for sulfate removal from the CDDS leachate. Good sulfate removal efficiencies (up to 99.9%) from the CDDS leachate can be achieved by precipitation with barium chloride ($BaCl_2$) and lead(II) nitrate ($Pb(NO_3)_2$). Precipitation with calcium chloride ($CaCl_2$), calcium carbonate ($CaCO_3$) and calcium oxide (CaO) gave less efficient sulfate removal. Adsorption of sulfate to aluminium oxide (Al_2O_3) yielded a 50% sulfate removal efficiency, whereas iron oxide coated sand (IOCS) as adsorbent gave only poor (10%) sulfate removal efficiencies.

7.1. Introduction

CDD originates from building, demolition and renovation of buildings and roads. Nearly 40% of the total mass of CDD is the fine fraction, called construction and demolition debris sand (CDDS), which consists of gypsum (52.4% of total gypsum) (Montero et al., 2010; Townsend et al., 2004). Reuse of this CDDS, which contains a high sulfate content, is a concern because of the chemical composition of the reused material and the potential risk to human health and the environment (Jang & Townsend, 2001). Therefore, limits have been set for the sulfate content of reused CDDS (1.73 g sulfate per kg of sand for The Netherlands).

Processes for sulfate removal from CDDS have been developed based on the leaching of the gypsum out from the CDDS material. Treatment of the CDDS leachate has been studied using biological sulfate reduction processes (Kijjanapanich et al., 2013a; Kijjanapanich et al., 2013b). A sulfate removal efficiency of 75-85% was achieved and the treated leachate can be reused in the CDDS leaching process (Kijjanapanich et al., 2013a; Kijjanapanich et al., 2013b). However, the biological sulfate reduction process has some disadvantages, including slow process kinetics, requirement and cost of an external electron donor and the need for a post-treatment of the sulfide containing CDDS leachate.

Removal of sulfate by chemical techniques can be an alternative to remove the sulfate contained in the CDDS leachate. Chemical precipitation is a widely used, proven technology for the removal of metals and other inorganics, suspended solids, fat, oils and greases from wastewater (U.S.EPA, 2000). Chemicals such as barium or calcium salts have been used for sulfate precipitation from mine water and academic laboratory waste chemicals (Benatti et al., 2009; Hlabela et al., 2007). The chemical precipitation processes require short treatment times, no need for a sophisticated operation and have low maintenance costs (requiring only replenishment of the chemicals used) (U.S.EPA, 2000) as compared to biological sulfate reduction processes.

This present study aims to develop a chemical removal process as an alternative for sulfate removal from CDDS leachate. Both precipitation and adsorption for sulfate removal from CDDS leachate were investigated to find an appropriate chemical sulfate removal process.

7.2. Material and Methods
7.2.1 Construction and demolition debris sand (CDDS) leachate

CDD samples were collected from Smink Afvalverwerking B.V. (Amersfoort, The Netherlands). Preparation of CDDS samples was accomplished as per the procedure in the study of Kijjanapanich et al. (2013a). CDDS was washed by demineralized water using a 1:10 ratio of CDDS:demineralized water at room temperature ($20 \pm 3°C$) until a constant sulfate concentration (around 1500 mg L^{-1}) was obtained in the leachate (approximately 2-3 d). The leachate was left for 1 d to allow the settling of the CDDS. The supernatant was then further used as the CDDS leachate for the experiments.

7.2.2 Experimental design

The experiments to study the effect of the chemical type, pH and the presence of calcium and acetate on the chemical sulfate removal can be divided into 4 steps (Table 7.1). First, a screening of chemicals to precipitate or adsorb the sulfate from the CDDS leachate was done at room temperature ($20 \pm 3°C$) with CDDS of a pH 7. The two chemicals which yielded the best sulfate removal (barium chloride ($BaCl_2$) and lead(II) nitrate ($Pb(NO_3)_2$)) were selected to study the effect of the initial CDDS leachate pH on the sulfate precipitation at room temperature ($20 \pm 3°C$) at different pH values (2, 5, 10 and 12). Hydrochloric acid (0.5 M) (HCl) and sodium hydroxide (0.5 M) (NaOH) solutions were used for pH adjustment. The precipitates from sulfate removal using $BaCl_2$ and $Pb(NO_3)_2$ at pH 7 were characterized based on their capillary suction time (CST), particle size distribution (PSD) and sludge volume index (SVI). The effect of calcium and acetate ions, which are contained in CDDS leachate (Kijjanapanich et al., 2013a), on sulfate precipitation were investigated at the third step.

7.2.3 Chemical sulfate precipitation

Jar tests were used to test the sulfate removal from CDDS leachate by chemical precipitation using $BaCl_2$, calcium chloride ($CaCl_2$), calcium carbonate ($CaCO_3$), calcium oxide (CaO) and $Pb(NO_3)_2$. In each jar test, CDDS leachate (500 ml) was filled in a 1 L beaker. All chemicals were supplied to the leachate 1.5 times the stoichiometric amount of the chemical precipitation reaction (Table 7.2). Then, the leachate was stirred at 200 rpm for 20 min. The leachate was then left for 1.5 h to investigate the appropriate settling time. During this 1.5 h, samples were collected at 15, 45 and 90 min, respectively. Each chemical was tested in triplicate.

7.2.4 Chemical sulfate adsorption

Jar tests were also used to test the sulfate removal from CDDS leachate by chemical adsorption using aluminium oxide (Al_2O_3) and iron oxide coated sand (IOCS). Al_2O_3 or IOCS was added in a 1:10 ratio (solid:liquid) in each jar test. The procedure of this test was same as describe in section 2.3 "Chemical Sulfate Precipitation".

Table 7.1. Conditions and parameters applied in each step of the experiments

Step	Chemicals	pH	Parameters		
			Sulfate (mg L^{-1})	Calcium (mg L^{-1})	Acetate (mg L^{-1})
I	Al$_2$O$_3$	7	1500	700	-
	BaCl$_2$	7	1500	700	-
	CaCl$_2$	7	1500	700	-
	CaCO$_3$	7	1500	700	-
	CaO	7	1500	700	-
	IOCS	7	1500	700	-
	Pb(NO$_3$)$_2$	7	1500	700	-
II	BaCl$_2$	2	1500	700	-
	BaCl$_2$	5	1500	700	-
	BaCl$_2$	10	1500	700	-
	BaCl$_2$	12	1500	700	-
	Pb(NO$_3$)$_2$	2	1500	700	-
	Pb(NO$_3$)$_2$	5	1500	700	-
	Pb(NO$_3$)$_2$	10	1500	700	-
	Pb(NO$_3$)$_2$	12	1500	700	-
III	Al$_2$O$_3$	7	1500	1000	-
	BaCl$_2$	7	1500	1000	-
	CaCl$_2$	7	1500	1000	-
	CaCO$_3$	7	1500	1000	-
	CaO	7	1500	1000	-
	IOCS	7	1500	1000	-
	Pb(NO$_3$)$_2$	7	1500	1000	-
	Al$_2$O$_3$	7	1500	700	1000
	BaCl$_2$	7	1500	700	1000
	CaCl$_2$	7	1500	700	1000
	CaCO$_3$	7	1500	700	1000
	CaO	7	1500	700	1000
	IOCS	7	1500	700	1000
	Pb(NO$_3$)$_2$	7	1500	700	1000
IV	BaCl$_2$	12	1500	700	-
	Pb(NO$_3$)$_2$	2	1500	700	-

7.2.5 Analytical methods

Sulfate removal was tested in VELP scientifica FC6S jar tests. The pH was measured using a micro pH 2001 pH meter and a 691 Metrohm pH meter using a SenTix 21 WTW pH electrode. Sulfate was measured with the turbidimetric method using a CECIL CE2030 UV visible spectrophotometer (Eaton et al., 2005), an Metrohm 883 Basic IC plus Ion Chromatography (IC) and an ICS-1000 Dionex IC (Eaton et al., 2005). Calcium was measured by the EDTA titration method and an AAnalyst 200 Perkin Elmer Atomic Absorption Spectrometer (AAS)-Flame (Eaton et al., 2005).

The dewatering properties of the precipitates were assessed by using a Triton CST Apparatus Model 200 (Triton Electronics Ltd., Essex, UK) with standard filter papers and an 18 mm sludge reservoir. PSD was calculated by DTS software (Malvern Instrument) using the dynamic light scattering method by a Zetasizer Nano ZS (Malvern Instrument) at a laser beam of 633 nm, a scattering angle of 173°, 23°C, refractive index of 1.64 and 1.89, and absorption of 0.440 and 0.184 at 633 nm for $BaSO_4$ and $PbSO_4$, respectively. SVI was measured using imhoff cones (Eaton et al., 2005).

Table 7.2. Stochiometry of the chemical sulfate precipitation reactions

Chemical	Reaction
$BaCl_2$	$Ba^{2+}_{(aq)} + SO_4^{2-}_{(aq)} \rightarrow BaSO_{4(s)}$
$CaCl_2$	$Ca^{2+}_{(aq)} + SO_4^{2-}_{(aq)} \rightarrow CaSO_{4(s)}$
$CaCO_3$	$Ca^{2+}_{(aq)} + SO_4^{2-}_{(aq)} \rightarrow CaSO_{4(s)}$
CaO	$Ca^{2+}_{(aq)} + SO_4^{2-}_{(aq)} \rightarrow CaSO_{4(s)}$
$Pb(NO_3)_2$	$Pb^{2+}_{(aq)} + SO_4^{2-}_{(aq)} \rightarrow PbSO_{4(s)}$

7.3. Results
7.3.1 Effect of chemicals on sulfate precipitation

Figure 7.1a shows the removal of sulfate using different chemicals. $BaCl_2$ and $Pb(NO_3)_2$ show good performance for sulfate precipitation (up to 99.9%), followed by CaO (30%) (Figure 7.1a). The initial sulfate concentration (1516 mg L^{-1}) was reduced to less than 2 mg L^{-1} in case of $BaCl_2$ (Table 7.3). From the calculations, around 8 mM or 1010 and 1660 mg L^{-1} of Ba^{2+} and Pb^{2+} remained in the system.

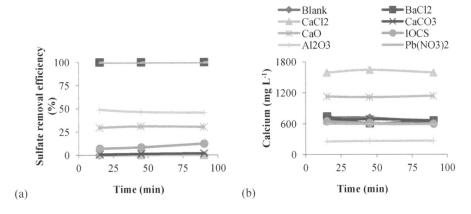

Figure 7.1. Performance of sulfate removal using jar tests as a function of operation time at pH 7: (a) Sulfate removal efficiency and (b) Calcium concentration.

Sulfate precipitation using $CaCl_2$ and $CaCO_3$ removed only around 3% of the sulfate (Figure 7.1a). There was an increase in the calcium concentration to 1620 and 1124 mg L^{-1} in the treated leachate when using $CaCl_2$ and CaO as chemical for sulfate precipitation, respectively (Figure 7.1b).

The pH of the initial CDDS leachate was around 7.3. The pH of the leachate remained at 7.0-7.6 after addition of $CaCl_2$, $CaCO_3$ and $BaCl_2$ for sulfate precipitation (Table 7.3). However, the pH of the CDDS leachate changed from 7.3 to 12.5 and 3.7, when adding CaO and $Pb(NO_3)_2$, respectively (Table 7.3). $BaCl_2$ and $Pb(NO_3)_2$ gave the best sulfate removal efficiency and were thus selected for the next step of the experiment.

Table 7.3. The effect of the chemical type on sulfate precipitation.

Chemicals	Initial pH	Final pH	Sulfate concentration (mg L^{-1})		Calcium concentration (mg L^{-1})	
			Initial	Final	Initial	Final
Precipitation						
BaCl$_2$	7.31 ± 0.01	7.52 ± 0.05	1516 ± 21	<2	694 ± 40	668 ± 55
CaCl$_2$	7.31 ± 0.01	7.35 ± 0.02	1516 ± 21	1504 ± 12	694 ± 40	1620 ± 40
CaCO$_3$	7.31 ± 0.01	7.48 ± 0.04	1516 ± 21	1500 ± 52	694 ± 40	668 ± 13
CaO	7.31 ± 0.01	12.51 ± 0.01	1516 ± 21	1055 ± 4	694 ± 40	1124 ± 13
Pb(NO$_3$)$_2$	7.31 ± 0.01	3.69 ± 0.1	1516 ± 21	3.7 ± 0.2	694 ± 40	594 ± 5
Adsorption						
Al$_2$O$_3$	7.31 ± 0.01	7.60 ± 0.03	1516 ± 21	827 ± 13	694 ± 40	262 ± 6
IOCS	7.31 ± 0.01	6.89 ± 0.03	1516 ± 21	1349 ± 2	694 ± 40	604 ± 35

7.3.2 Effect of initial CDDS leachate pH on sulfate precipitation

The effect of pH on sulfate precipitation was investigated using BaCl$_2$ and Pb(NO$_3$)$_2$ at room temperature (20 ± 3°C).

7.3.2.1 Barium chloride (BaCl$_2$)

The sulfate removal efficiencies did not change significantly at different pH values (pH 2, 5, 7, 10 and 12): sulfate removal efficiencies of 99.87-99.92% were achieved (Figure 7.2a). The highest sulfate removal efficiency (99.92%) was achieved at pH 7. The calcium concentration did not change significantly (Figure 7.2b).

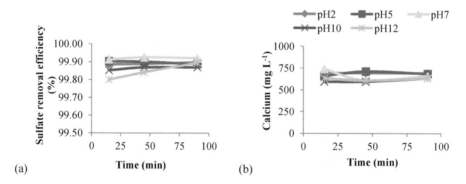

(a)   (b)

Figure 7.2. Performance of sulfate precipitation using BaCl$_2$ as a function of operation time at different pH: (a) Sulfate removal efficiency and (b) Calcium concentration.

7.3.2.2 Lead(II) nitrate (Pb(NO$_3$)$_2$)

The sulfate removal efficiencies vary between 98.50 and 99.90% when using Pb(NO$_3$)$_2$. CDDS leachate at pH 5 and 7 yielded the best sulfate removal efficiency of 99.90% (Figure 7.3a). The lowest sulfate removal efficiency (98.50%) was achieved with CDDS leachate at pH 12. The calcium concentrations of the CDDS leachate did not change at all pH values investigated (Figure 7.3b).

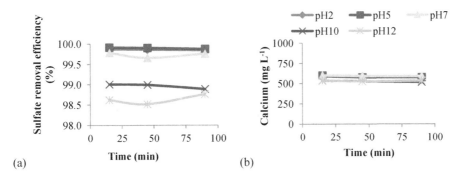

Figure 7.3. Performance of sulfate precipitation using Pb(NO$_3$)$_2$ as a function of operation time at different pH: (a) Sulfate removal efficiency and (b) Calcium concentration.

7.3.3 Precipitate characterization

XRD and Visual MINTEQ software analysis showed that the precipitate from sulfate precipitation using BaCl$_2$ and Pb(NO$_3$)$_2$ were solely barite (BaSO$_4$) and anglesite (PbSO$_4$) (data not shown). The CST of the precipitates using BaCl$_2$ and Pb(NO$_3$)$_2$ were 6.0 and 6.2 s with a TSS of 9.6 and 11.2 g L^{-1}, respectively. Both precipitates showed good settling properties with a SVI of 5.2 and 1.7 mL g^{-1} at a TSS of 3 g L^{-1} of the BaSO$_4$ and PbSO$_4$, respectively. The average PSD of the precipitates was bigger than 3.5 μm in both BaCl$_2$ and Pb(NO$_3$)$_2$ test (data not shown).

7.3.4 Effect of chemicals on sulfate adsorption

A sulfate removal efficiency of 50% was achieved when using Al$_2$O$_3$ as adsorbent for sulfate removal, while IOCS gave only a poor sulfate removal efficiency (10%) (Figure 7.1a). The sulfate removal efficiency of Al$_2$O$_3$ was reduced to only 10% in the absence of calcium ions (Figure 7.4a).

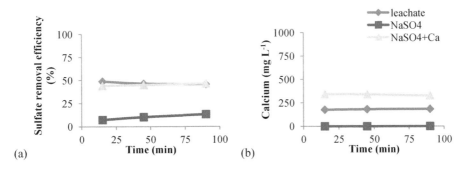

Figure 7.4. The effect of calcium on the performance of sulfate precipitation using Al$_2$O$_3$ as a function of operation time: (a) Sulfate removal efficiency and (b) Calcium concentration.

7.3.5 Effect of calcium and acetate on sulfate removal

Calcium had no effect on sulfate removal with $BaCl_2$, $CaCl_2$, $CaCO_3$, CaO, $Pb(NO_3)_2$ and IOCS (data not shown). However, calcium highly affected the sulfate removal in case of Al_2O_3 (Figure 7.4a). A sulfate removal efficiency of approximately 10% was achieved when Al_2O_3 was used with a sodium sulfate (Na_2SO_4) solution (in the absence of calcium). However, the sulfate removal efficiency increased from 10% to 50% in case of CDDS leachate and a Na_2SO_4 solution mixed with $CaCl_2$ (Figure 7.4a). In case of acetate, there was no effect of acetate on the sulfate precipitation or adsorption with all chemicals tested in this experiment (data not shown).

7.4. Discussion
7.4.1 Physico-chemical methods for sulfate removal

This study showed that the highest sulfate removal efficiency was achieved with $BaCl_2$ or $Pb(NO_3)_2$ as chemical for sulfate precipitation at pH 7. This is due to the low solubility of 2.66 (Benatti et al., 2009) and 38.40 mg L^{-1} (Benatti et al., 2009) at 25 °C of $BaSO_4$ and $PbSO_4$, respectively. The solubility of $BaSO_4$ and $PbSO_4$ are extremely low as compared with the solubility of gypsum ($CaSO_4$), which has a solubility of 2600 mg L^{-1} in pure water at 25 °C (FAO, 1990). The latter results in a residual sulfate concentration of 1450 mg L^{-1} if calcium was used as precipitant. The effect of the pH (Figures 7.2 and 7.3) on sulfate precipitation using $BaCl_2$ and $Pb(NO_3)_2$ can be negligible due to the low solubility of $BaCl_2$ and $Pb(NO_3)_2$ (Benatti et al., 2009). However, the final pH of the treated CDDS leachate becomes acid (pH 3.7) when using $Pb(NO_3)_2$, while the final pH is still neutral (pH 7.5) in case of $BaCl_2$ (Table 7.3). Precipitation with Ba^{2+} or Pb^{2+} is thus potentially an effective method for complete conversion of dissolved sulfate to an insoluble form, which is mainly barite and anglesite (data not shown). However, the major disadvantages of precipitation with either Ba or Pb are the handling of the toxic compounds ($BaCl_2$ and $Pb(NO_3)_2$) to be added to the CDDS leachate, the generation of Ba^{2+} or Pb^{2+} containing waste that requires disposal and the need for a post-treatment of the CDDS leachate to remove the remaining dissolved Ba and Pb (Benatti et al., 2009).

Table 7.4 shows that the sulfate removal efficiency depends on many parameters, such as initial sulfate concentration and pH. Table 7.4 compares studies of chemical sulfate precipitation using calcium, barium and lead salts. The precipitation of sulfate with calcium is another option, which has no toxic risks and produces gypsum that can be a replacement for natural gypsum (Lens et al., 1998). Benatti et al. (2009) found that $CaCl_2$ showed a good sulfate precipitation performance (>99%) at pH 4.0 with waste chemicals from academic laboratories. In contrast, no sulfate precipitation was observed when using $CaCl_2$ in this present study. This was mainly due to the initial sulfate concentration used in this study (1516 mg L^{-1}) which was near gypsum solubility and much lower than those used in Benatti et al. (2009)'s study (142000-151000 mg L^{-1}). An alternative source of calcium as $CaCO_3$ was also tested. Figure 7.1b shows that $CaCO_3$ did not dissolve (leachate calcium concentration did not increase) due to its low solubility, resulting in almost no sulfate precipitation (3% removal efficiency).

The IOCS shows good performance in many studies for the removal of arsenic (Petrusevski et al., 2007; Thirunavukkarasu et al., 2001; Vaishya & Gupta, 2006; Yuan et al., 2002). However, IOCS gave only a poor sulfate removal efficiency (10%). Al_2O_3

is widely used for adsorption of many compounds such as phosphate and metals (Genz et al., 2004; Pavlova & Sigg, 1988; Tanada et al., 2003). In this study, the highest sulfate removal efficiency of 50% was achieved when using Al_2O_3 as adsorbent. This is higher than those found in the study of Kawasaki et al. (2008) where the highest sulfate removal efficiency of only 5.8% at pH 9.7 using calcined aluminum oxide. This is because of the presence of phosphate in the tested solution (Kawasaki et al., 2008), which can adsorbs better on Al_2O_3 than sulfate. In contrast, the higher sulfate removal efficiency might be due to the interference of calcium ions which are present in system of this present study.

Table 7.4. Chemical sulfate precipitation using calcium and barium salts.

Chemicals	pH	Sulfate concentration (mg L^{-1})		Removal efficiency (%)	References
		Initial	Final		
CaO	9.3	2060	1970	4	Bosman et al.
BaS	12.0	1970	120	94	(1990)
CaO	12.0	2650	1250	53	Maree et al.
BaS	11.9	1250	250	80	(2004)
CaO	10.0	2275	2000	12	Hlabela et al.
BaCO$_3$	10.0	2000	200	90	(2007)
CaCl$_2$	4.0	142000-151000	1000	>99	Benatti et al.
BaCl$_2$	4.0	142000-151000	-	52-61	(2009)
CaCl$_2$	7.3	1516	1504	3	This study
CaCO$_3$	7.5	1516	1500	3	
CaO	12.5	1516	1055	32	
BaCl$_2$	2.0	1516	<2	>99	
	5.0	1516	<2	>99	
	7.5	1516	<2	>99	
	10.0	1516	2.0	>99	
	12.0	1516	3.0	>99	
Pb(NO$_3$)$_2$	2.0	1516	2.0	>99	
	5.0	1516	<2	>99	
	7.5	1516	3.7	>99	
	10.0	1516	16	99	
	12.0	1516	21	98	

7.4.2 Sulfate precipitation for CDDS leachate treatment

Calcium and acetate contained in the CDDS leachate did not show any significant effect on the sulfate removal efficiency using $BaCl_2$ or $Pb(NO_3)_2$ precipitation. This due to the low solubility of $BaCl_2$ and $Pb(NO_3)_2$. It was confirmed by XRD and Visual MINTEQ software that the precipitate from sulfate precipitation using $BaCl_2$ and $Pb(NO_3)_2$ were solely barite ($BaSO_4$) and anglesite ($PbSO_4$) (data not shown). Barium carbonate ($BaCO_3$) and barium sulfide can be alternative chemicals for sulfate precipitation (Bosman et al., 1990; Hlabela et al., 2007; Maree et al., 2004). $BaCO_3$ can only be used for the removal of sulfate from wastewater that also contains a lot of calcium. This is because calcium is required to remove the carbonate. This chemical is nevertheless not suitable for metal containing leachate treatment, because $BaCO_3$ becomes inactive when coated with metal hydroxide precipitates (Maree et al., 2004). Moreover, a problem in separating $BaSO_4$ and $CaCO_3$, which co-precipitate, has to be overcome (Maree et al., 2004). In case of BaS, a high sulfate removal efficiency can be achieved (Maree et al.,

2004). However, a sulfide trapping and sulfur recovery unit are required when using BaS.

Calcium was found to affect the sulfate sorption onto Al_2O_3. It is possible that the sulfate concentration near the Al_2O_3 surface exceeds the sulfate concentration in the CDDS leachate, due to the adsorption of sulfate. When sulfate is continuously adsorbed, calcium contained in the system can precipitate with this adsorbed sulfate as calcium sulfate (gypsum) and attach to Al_2O_3, resulting in reducing both the sulfate and calcium concentration in the system.

7.4.3 Chemical versus biological treatment of CDDS leachate for sulfate removal

Table 7.5 compares the use of chemical and biological sulfate removal processes for CDDS leachate treatment. Chemical precipitation is a well-established technology with ready availability of equipment and many chemicals (U.S.EPA, 2000). Chemical sulfate removal processes require a short time for treatment (minutes time scale) and a low maintenance as compared to biological sulfate reduction processes (hours or days time scale). Therefore, a small reactor volume is required for the chemical sulfate precipitation process. Moreover, chemical sulfate precipitation requires only replenishment of the chemical used. In contrast, continuous supply of electron donor is required in case of biological sulfate reduction and the H_2S is a product which requires a post-treatment.

Table 7.5. Comparison between chemical and biological sulfate removal technologies from CDDS leachate treatment.

Parameter	Chemical sulfate precipitation	Biological sulfate reduction and sulfur recovery
Sulfate removal	Direct removal	Convert sulfate to sulfide and sulfide oxidation
Time	Fast	Slow
Reactor size	Small	Large
Product	Sulfate precipitate	H_2S and elemental sulfur
Electron donor	No need	Required
Chemical needed	Ba^{2+} or Pb^{2+} (expensive)	Ethanol (sulfate reduction), Oxygen (sulfide oxidation)
Sludge	Chemical sludge	Elemental sulfur and bio-anaerobic sludge

The chemicals used in chemical precipitation processes can nevertheless be expensive. Besides, although $BaCl_2$ and $Pb(NO_3)_2$ show good performance in sulfate precipitation, Ba^{2+} and Pb^{2+}, which remained in the leachate (8 mM) after the precipitation process, are toxic compounds. They can result in an adverse impact on the environment if the leachate is directly discharged without any post-treatment. The minimum amount of $BaCl_2$ and $Pb(NO_3)_2$ required for sulfate removal needs to be investigated in order to minimize the amount of chemical used and reduce the remaining toxic compounds in the treated water. Moreover, systems for precipitate separation and appropriate reuse or disposal of the solid phase are necessary (Silva et al., 2002). For example, $BaSO_4$ can be converted to BaS, due to reducing conditions created by the conversion of coal to CO and CO_2, using a Muffle furnace as reported in Maree et al. (2004).

Recent research concentrates on combining chemical precipitation with other treatment methods such as photochemical oxidation (U.S.EPA, 2000), reverse osmosis (RO), membrane extraction and ion exchange resins (Guimarães & Leão, 2011; Kratochvil et al., 2008; Simkin et al., 2004) to optimize performance. The combination of chemical sulfate precipitation with membrane technology can be an attractive option to separate toxic compounds used for sulfate precipitation from the CDDS leachate or treated wastewater (Figure 7.5). The membrane process can be used either before or during the precipitation process. The membrane process, such as RO or an anion exchange membrane, can be used for separation of sulfate from the CDDS leachate. The sulfate contained in the retentate is then precipitated with the chemical either in the same unit or separately in a crystallization unit. The sludge produced from such a process is easy to manage due to the more concentrated starting sulfate concentration.

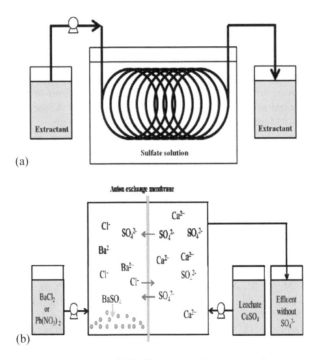

Figure 7.5. Schematic diagrams of (a) silicone membrane extraction reactor and (b) anion exchange membrane reactor.

7.5. Conclusions

This study demonstrated the feasibility of chemical sulfate precipitation and adsorption for sulfate removal from CDDS leachate using various chemicals, such as barium, lead and aluminium salts. $BaCl_2$ and $Pb(NO_3)_2$ yielded a high sulfate precipitation efficiency (up to 99.9%). The effect of the initial CDDS leachate pH on sulfate precipitation using $BaCl_2$ and $Pb(NO_3)_2$ can be negligible. However, Ba^{2+} and Pb^{2+} are toxic compounds, further research is thus needed to investigate new separation technologies for sulfate precipitation to minimize their use or to explore the use of other non-toxic chemicals with a low solubility product of the sulfate salt.

7.6 References

Benatti, C. T., Tavares, C. R. G., & Lenzi, E. (2009). Sulfate removal from waste chemicals by precipitation. *J. Environ. Manage., 90*, 504-511.

Bosman, D. J., Clayton, J. A., Maree, J. P., & Adlem, C. J. L. (1990). Removal of sulphate from mine water with barium sulphide. *Mine Water Environ., 9*, 149-163.

Eaton, A. D., APHA, AWWA, & WEF. (2005). *Standard methods for the examination of water and wastewater* (21st ed.). Washington D.C.

FAO. (1990). *FAO Soils Bulletin 62: Management of gypsiferous soils*. Rome.

Genz, A., Kornmüller, A., & Jekel, M. (2004). Advanced phosphorus removal from membrane filtrates by adsorption on activated aluminium oxide and granulated ferric hydroxide. *Water Res., 38*(16), 3523-3530.

Guimarães, D., & Leão, V. A. (2011). *Studies of sulfate ions removal by the polyacrylic anion exchange resin Amberlite IRA458: batch and fixed-bed column studies.* Paper presented at the 11th International Mine Water Association Congress – Mine Water – Managing the Challenges, Aachen, Germany.

Hlabela, P., Maree, J., & Bruinsma, D. (2007). Barium carbonate process for sulphate and metal removal from mine water. *Mine Water Environ., 26*, 14-22.

Jang, Y. C., & Townsend, T. (2001). Sulfate leaching from recovered construction and demolition debris fines. *Adv. Environ. Res., 5*, 203-217.

Kawasaki, N., Ogata, F., Takahashi, K., Kabayama, M., Kakehi, K., & Tanada, S. (2008). Relationship between anion adsorption and physicochemical properties of aluminum oxide. *J. Health Sci., 54*(3), 324-329.

Kijjanapanich, P., Annachhatre, A. P., Esposito, G., van Hullebusch, E. D., & Lens, P. N. L. (2013a). Biological sulfate removal from gypsum contaminated construction and demolition debris. *J. Environ. Manage., 131*, 82-91.

Kijjanapanich, P., Do, A. T., Annachhatre, A. P., Esposito, G., Yeh, D. H., & Lens, P. N. L. (2013b). Biological sulfate removal from construction and demolition debris leachate: Effect of bioreactor configuration. *J. Hazard. Mater., In Press*.

Kratochvil, D., Marchant, B., Bratty, M., & Lawrence, R. (2008). *Innovation in ion-exchange technology for the removal of sulfate.* Paper presented at the the 69th Annual International Water Conference, San Antonio, USA.

Lens, P. N. L., Visser, A., Janssen, A. J. H., Hulshoff Pol, L. W., & Lettinga, G. (1998). Biotechnological treatment of sulfate rich wastewaters. *Crit. Rev. Environ. Sci. Technol., 28*(1), 41-88.

Maree, J. P., Hlabela, P., Nengovhela, R., Geldenhuys, A. J., Mbhele, N., Nevhulaudzi, T., et al. (2004). Treatment of mine water for sulphate and metal removal using barium sulphide. *Mine Water Environ., 23*, 195-203.

Montero, A., Tojo, Y., Matsuto, T., Yamada, M., Asakura, H., & Ono, Y. (2010). Gypsum and organic matter distribution in a mixed construction and demolition waste sorting process and their possible removal from outputs. *J. Hazard. Mater., 175*, 747-753.

Pavlova, V., & Sigg, L. (1988). Adsorption of trace metals on alumimium oxide: A similation of processes in freahwater systems by close approximation to natural conditions. *Water Res., 22*(12), 1571-1575.

Petrusevski, B., Sharma, S., Schippers, J. C., & Shordt, K. (2007). *Arsenic in Drinking Water*. Delft, The Netherlands: IRC International Water and Sanitation Centre.

Silva, A. J., Varesche, M. B., Foresti, E., & Zaiat, M. (2002). Sulphate removal from industrial wastewater using a packed-bed anaerobic reactor. *Process Biochem., 37*, 927-935.

Simkin, S. M., Lewis, D. N., Weathers, K. C., Lovett, G. M., & Schwarz, K. (2004). Determination of sulfate, nitrate, and chloride in throughfall using ion-exchange resins. *Water Air Soil Pollut., 153*, 343-354.

Tanada, S., Kabayama, M., Kawasaki, N., Sakiyama, T., Nakamura, T., Araki, M., et al. (2003). Romoval of phosphate by aluminum oxide hydroxide. *J. Colloid Interface Sci., 257*(1), 135-140.

Thirunavukkarasu, O. S., Viraraghavan, T., & Subramanian, K. S. (2001). Removal of arsenic in drinking water by iron oxide-coated sand and ferrihydrite - Batch studies. *Water Qual. Res. J. Canada, 36*(1), 55-70.

Townsend, T., Tolaymat, T., Leo, K., & Jambeck, J. (2004). Heavy metals in recovered fines from construction and demolition debris recycling facilities in Florida. *Sci. Total Environ., 332*, 1-11.

U.S.EPA. (2000). *Wastewater Technology Fact Sheet Chemical Precipitation.* Washington, D.C.

Vaishya, R. C., & Gupta, S. K. (2006). Arsenic(V) removal by sulfate modified iron oxide-coated sand (SMIOCS) in a fixed bed column. *Water Qual. Res. J. Canada, 41*(2), 157-163.

Yuan, T., Hu, J. Y., Ong, S. L., Luo, Q. F., & Ng, W. J. (2002). Arsenic removal from household drinking water by adsorption. *J. Environ. Sci. Health, Part A: Toxic/Hazard. Subst. Environ. Eng., 37*(9), 1721-1736.

CHAPTER 8

Spontaneous Electrochemical Treatment for Sulfur Recovery by a Sulfide Oxidation/ Vanadium(V) Reduction Galvanic Cell

This chapter was submitted for publication as:
Kijjanapanich, P., Kijjanapanich, P., Annachhatre, A. P., Esposito, G., & Lens, P. N. L. (2013). Spontaneous electrochemical treatment for sulfur recovery by a sulfide oxidation/vanadium(V) reduction galvanic cell. *J. Electroanal. Chem., Submitted.*

Chapter 8

Sulfide is the product of the biological sulfate reduction process which gives toxicity and odor problems. Wastewater or bioreactor effluent containing sulfide can cause several environmental impacts. Therefore, the removal of sulfide from the effluent of biological sulfate reducing reactors or wastewater is necessary. Electrochemical treatment is one of the alternatives for sulfide removal and sulfur recovery from such aqueous sulfide containing solutions. This study aims to develop a spontaneous electrochemical sulfide oxidation/vanadium(V) reduction cell with a graphite electrode system to recover sulfide as elemental sulfur. A high surface area of the graphite electrode is required in order to have as less internal resistance as possible. A sulfide removal efficiency up to 91% was achieved when using five graphite rods with powder graphite as electrode at an external resistance of 30 Ω and a sulfide concentration of 250 mg L^{-1}.

8.1. Introduction

Sulfide can be found in many domestic and industrial wastewaters (Dutta et al., 2008; Pikaar et al., 2012; Pikaar et al., 2011) as well as in the effluent of sulfate reducing bioreactors (Kijjanapanich et al., 2013a; Kijjanapanich et al., 2013b). This sulfide not only yields offensive odor problems, but also introduces toxicity and sewer pipe corrosion (Vincke et al., 2001).

Sulfide can also interfere with the iron-phosphate precipitates in soils and sediments due to the formation of iron sulfides and associated release of phosphorous. Exhaustion of iron from iron sulfide precipitates results in increased sulfide levels and iron shortage, while the increased phosphate mobilization and the disturbed iron cycle result in increased phosphate levels in the water phase (Smolders & Roelofs, 1993). In this way, the released phosphate causes indirect eutrophication resulting, among others, in a dominance of non-rooting species and algae, and thus increased turbidity of the water (Smolders & Roelofs, 1993). Therefore, sulfide removal from wastewater or the effluent of biological sulfate reducing reactors prior to discharge into the environment is required from both an environmental and economic point of view (Dutta et al., 2008).

Precipitation of sulfide as metal sulfide, particularly iron sulfide (Firer et al., 2008; Nielsen et al., 2008; Zhang et al., 2009) is a common sulfide removal process. Sulfide oxidation to elemental sulfur (S^0) is an alternative (Lens et al., 2002; Sahinkaya et al., 2011), which offers several advantages over the aforementioned method (González-Sánchez & Revah, 2009).

Conversion of sulfide to elemental sulfur either in acid or base conditions is an oxidation reaction (Equations 8.1 and 8.2) where an electron acceptor is required to fulfill the redox reaction. Either chemical or biological processes can be applied for sulfide oxidation to elemental sulfur (González-Sánchez & Revah, 2007). Nowadays, biological sulfide oxidation using oxygen as electron acceptor and sulfide oxidizing bacteria as a catalyst is a very popular system (González-Sánchez & Revah, 2009; Henshaw & Zhu, 2001; Krishnakumar et al., 2005; Sahinkaya et al., 2011). However, this system requires energy for oxygen supply (Syed et al., 2006; van den Ende et al., 1996), complicated operation techniques (Syed et al., 2006) and the pH conditions of these biological systems are usually mildly or extremely acidic (Gabriel & Deshusses,

2003; Kraakman, 2003). Oversupply of oxygen also yields low sulfide removal efficiencies since most sulfide is changed to sulfate instead of elemental sulfur (Janssen et al., 1995).

Acid solution: $H_2S_{(g)} \rightarrow S_{(s)} + 2H^+{}_{(aq)} + 2e^-$ $\qquad\qquad$ (8.1)

Base solution: $S^{2-}{}_{(aq)} \rightarrow S_{(s)} + 2e^-$ $\qquad\qquad$ (8.2)

Electrochemical treatment of such wastewaters can be an appropriate way that offers several advantages, including good energetic efficiency, environmental compatibility, versatility, selectivity and cost effectiveness (Ángela et al., 2009; Dutta et al., 2009; Rajeshwar et al., 1994). The ideal electron acceptor is the one which can provide a spontaneous reaction or produce a galvanic cell. Not only the sulfide can be removed, but the elemental sulfur can be recovered and electricity will also be generated when the oxidation and reduction reactions occur in separated chambers.

By this principle, some galvanic cells have been developed for treating sulfide containing wastewater. One of them used hexacyanoferrate (III) ion ($Fe(CN)_6^{3-}$) as an electron acceptor (Dutta et al., 2008; Dutta et al., 2009). However, in this study, vanadium with oxidation state 5+, i.e., VO_2^+ is selected, as it has been already thermodynamically shown that VO_2^+ is able to perform a spontaneous redox reaction with sulfide/sulfur oxidation as illustrated in Equation 8.3 and 8.4. Referring to these equations, the sulfide presented in ion form (S^{2-}) gives a higher standard cell potential ($E^0{}_{cell}$) than those of H_2S. This means that electricity generated by alkaline sulfide wastewater treatment cells should be higher than by acidic cells.

$H_2S_{(g)} + 2VO_2^+{}_{(aq)} + 2H^+{}_{(aq)} \rightarrow S_{(s)} + 2VO^{2+}{}_{(aq)} + 2H_2O \qquad E^\circ_{cell} = +0.86 \ V$ $\quad$ (8.3)

$S^{2-}{}_{(aq)} + 2VO_2^+{}_{(aq)} + 4H^+{}_{(aq)} \rightarrow S_{(s)} + 2VO^{2+}{}_{(aq)} + 2H_2O \qquad E^\circ_{cell} = +1.48 \ V$ $\quad$ (8.4)

Whenever oxidation and reduction chambers are connected with an external resistance (R) (Figure 8.1), electrons will transfer from the oxidation (anode) to the reduction (cathode) parts. Thus, direct electric current (I) occurs. Moreover, the amount of sulfide changed to elemental sulfur varies in accordance with the amount of electron flow through the cell circuit. From Equation 8.1, production of one mole of elemental sulfur, two moles of electrons have to be transferred. This means that the rate of electron transfer, i.e., electric current, determines the rate of sulfide reduction or elemental sulfur production. Therefore, the maximum electric current provides the maximum sulfide removal efficiency.

Theoretically, such current production depends on both the external and internal cell resistance (r) as shown in Equation 8.5. The highest current production will be obtained when minimum resistances, both external and internal, are employed. There are many factors affecting the internal cell resistance, for example, type and surface area of the electrode, surface area of the cation exchange membrane, concentration of ions in the solution, etc. If other factors are fixed, electrical current will depend directly on the internal cell resistance.

$$I = \frac{E_{cell}}{R+r} \qquad\qquad (8.5)$$

Many thermodynamic spontaneous reactions proceed at very slow rates at ambient temperature and pressure. Pre-testing prior to this study with mixing of a sulfide solution with a metavanadate solution showed that yellow precipitates of elemental sulfur occur immediately (data not shown). This shows that the redox reaction is a spontaneous reaction both from a thermodynamic and kinetic point of view.

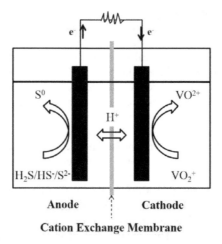

Figure 8.1. Schematic representation of the electrochemical sulfide oxidation/vanadium(V) reduction reactor

Therefore, this research was conducted to treat the effluent of sulfate reducing bioreactors by using a spontaneous electrochemical sulfide oxidation/vanadium(V) reduction in a graphite electrode system. Performance of the sulfide removal efficiency at different pH values was evaluated. The effect of the internal resistance on the removal efficiency and electrical current was also investigated.

8.2. Material and Methods
8.2.1 Sulfide wastewater samples
8.2.1.1 Synthetic sulfide wastewater

The synthetic wastewater used in this study was sulfide in buffer solutions of pH 7 and 10. The pH 10 buffer solution was prepared by dissolving a carbonate buffer, i.e., 5 g $NaHCO_3$ + 1 g NaOH in anoxic water (boiled and cooled to ambient temperature demineralized water), then 1872 mg washed crystals of sodium sulfide ($Na_2S \cdot 9H_2O$) were added and dissolved. The final volume was made up to 1 L with anoxic water. The concentration of this sulfide solution was 7.8 mM, corresponding to 250 mg L^{-1} of sulfide. As for sulfide in a pH 7 buffer solution, phosphate buffer, i.e., 4 g Na_2HPO_4 + 5 g KH_2PO_4 + 1 g NaCl were dissolved instead of the carbonate buffer. The final pH of both solutions were 11 and 8, respectively.

8.2.1.2 Real sulfide wastewater sample

The effluent of a sulfate reducing bioreactor treating construction and demolition debris (CDD) leachate described by Kijjanapanich et al. (2013c) was used as a real effluent for the electrochemical treatment of real wastewater samples.

8.2.2 Vanadium solution

A vanadium solution (70 mM) was prepared by dissolving 4 g of ammonium metavanadate (NH$_4$VO$_3$) with 1 M NaOH 100 mL. The solution was warmed until NH$_4$VO$_3$ totally dissolved. Then, 160 mL of 1 M H$_2$SO$_4$ was carefully added. The solution was cooled down to ambient temperature and the volume was made up to 500 mL by deminerized water.

8.2.3 Electrochemical sulfide oxidation/vanadium(V) reduction reactors

Abiotic electrochemical reactors made of polypropylene (PP) (Figure 8.2) were used as galvanic cells. Each of them had a working volume of 1 L chamber. This chamber was divided equally into two parts by a cation exchange membrane (Ultrex, CM17000, Membranes International Inc., USA). The anode part was filled with 400 mL sulfide wastewater sample, while the other part named cathode was filled with the same volume of vanadium solution. A numbers of graphite rods (5.6 mm diameter and 80 mm long) (GIOCONDA 6, KOH-I-NOOR HARDTMUTH a.s., Czech Republic) acted as electrodes. These were immersed in both anode and cathode compartments. The anode and cathode electrodes of each cell were connected with an external resistance.

8.2.4 Electrochemical sulfide oxidation/vanadium(V) reduction experiments
8.2.4.1 Experiments with synthetic sulfide wastewater

The experiments with synthetic sulfide wastewater comprised of three parts in series as followed. All experiments were carried out in triplicate.

8.2.4.1.1 Determination of internal cell resistance

Determination of the internal cell resistance was carried out by using three laboratory galvanic cells with different numbers and types of electrodes, i.e., single rod, five rods and five rods with powder graphite (five plus). In the five plus cell, grinded graphite powder (162 g and 150 mL) of $\leq$ 0.5 mm in size was added. The synthetic sulfide solution with a buffer of pH 10 was used as an anode solution. The external resistances connected between the anode and cathode electrodes of each cell were varied from 5.6 to 1500 Ω. The electric current was measured immediately after connecting with an external resistance. Then, the internal resistance was approximated by plotting a graph between the fraction of 1/I values and the external resistances.

Equation 8.5 can be rewritten as shown in Equation 8.6. A plot of 1/I versus R thus gives a straight line with a slope of 1/E$_{cell}$ and an intercept of r/E$_{cell}$. Then E$_{cell}$ and r can be calculated by this slope and interception.

$$\frac{1}{I} = \frac{R}{E_{cell}} + \frac{r}{E_{cell}}$$
(8.6)

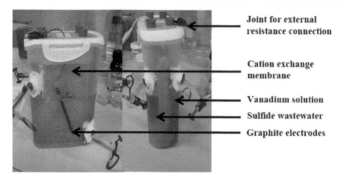

Joint for external
resistance connection

Cation exchange
membrane

Vanadium solution
Sulfide wastewater
Graphite electrodes

Figure 8.2. Lab-scale photograph of the electrochemical sulfide oxidation/vanadium(V) reduction reactor.

8.2.4.1.2 Comparison of sulfide removal efficiencies at different external and internal cell resistances

Batch experiments were conducted by using the same three different internal resistance cells as mentioned in 2.4.1.1. A synthetic sulfide solution with buffer at pH 10 was also used as anode solution. Three different external resistances of 30, 70 and 140 Ω were applied. The remaining sulfide concentration and pH value at different operation times were measured. Electric current and voltage across the electrodes were also recorded.

8.2.4.1.3 Investigation of the pH effect on the sulfide removal efficiency

Batch experiments were performed by using the minimum internal resistance cells, i.e., the five plus cell. Sulfide removal efficiencies at the same three different external resistances as mentioned in 2.4.1.2 were compared. The effect of the pH was investigated with the synthetic sulfide solution of both a pH 7 and 10 buffer.

8.2.4.2 Experiments with real effluent

Batch experiments were performed by using the same minimum internal resistance cells as mentioned in 2.4.1.3. The remaining sulfide concentration and pH value at different operation times were measured at a minimum external resistance of 30 Ω. The effluent of the sulfate reducing bioreactor (Kijjanapanich et al., 2013c), treating CDD leachate at its original pH (8.5) and adjusted to pH 10 were supplied as anode solutions.

8.2.5 Analytical methods

The pH was measured using a 691 Metrohm pH meter and a SenTix 21 WTW pH electrode. Voltage and resistance were measured using a Klaasing Electronics METEX M-4650 digital multimeter. Sulfide was measured by the method of Ralf Cord-Ruwisch (Ruwisch, 1985) using a Perkin Elmer Lambda 20 UV visible spectrophotometer. Sulfate was measured using an ICS-1000 Dionex Ion Chromatography (IC) (Eaton et al., 2005). Thiosulfate was measured by iodometric titration (Eaton et al., 2005).

8.3. Results

8.3.1 Internal resistance of the galvanic cells with different types of graphite electrodes

Figure 8.3a shows the variation of the average electric currents at different external resistance values of each cell at pH 10. Data fluctuations among the triplicate tests are also demonstrated (Figure 8.3a), these deviations became smaller at high voltage values. About 50% of triplicate tests gave not more than 5 percent relative deviation (%RSD), while 83% gave not more than 10%RSD. When 1/I was plotted versus R as illustrated in Figure 8.3b, results were obtained as demonstrated in Table 8.1. The internal cell resistance reduced when the numbers of graphite electrodes increased. The internal resistance of 1, 5 graphite rods and five plus cells were 1114, 400 and 58 Ω, respectively. Each calculated or measured cell potential of both single and 5 rods cells were equal while those of the five plus cell was higher. The highest current (15 mA) and highest potential as well as the lowest internal resistance (58 Ω) were achieved with the five plus cell.

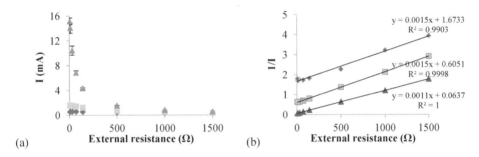

(a) (b)

Figure 8.3. Evolution of the current (I) and 1/I with different external resistance values: (a) the current and (b) 1/I. ($\blacklozenge$) 1 graphite rod, ($\blacksquare$) 5 graphite rods and ($\blacktriangle$) five plus electrodes.

Table 8.1. Internal cell resistance determined by plotting 1/I versus R

Type of electrodes	Contact area (cm^2)	Obtained relation, (R^2)	Calculated E_{cell} (mV)	Measured voltage (mV)	Approximated internal resistance (Ω)
Single rod	14.1	1/I =0.0015R+1.67 (0.99)	667	769	1114
Five rods	70.4	1/I =0.0015R+0.60 (1.0)	667	771	400
Five plus	>633	1/I =0.0011R+0.064 (1.0)	909	878	58

8.3.2 Performance of galvanic cells at pH 10 at different external and internal cell resistance

Voltage, electrical current and the remaining sulfide concentrations at different operating times and external resistances of each cell at pH 10 are illustrated in Figures 8.4 and 8.5-left. The voltage values were related with the external and internal resistances in the opposite way. At any internal resistance and operating time, a higher voltage was achieved when a higher external resistance was applied. In contrast, reduced internal resistance enhanced cell voltage. However, when the internal resistance was comparable high as for the single rod cell, the voltage values at any external resistance used were not significantly different. The measured pH of the anode solution gradually reduced from 11 to 8 over the 24 h operation time.

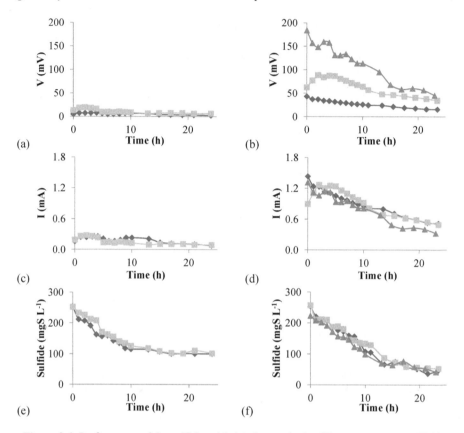

Figure 8.4. Performance of the sulfide oxidation for synthetic effluent treatment at pH 10 as a function of operation time (Left: 1 graphite rods and Right: 5 graphite rods electrodes): (a, b) voltage, (c, d) current and (e, f) sulfide. (♦) 30, (■) 70 and (▲) 140 Ω.

In all experiments, sulfide concentrations decreased simultaneously over the operating time during the initial period. Then, they reached a rather constant level. The operating times, which required for the sulfide contents to be constant, named exhausted time (Klymenko & Kulys, 2008), depended mainly on the type of cell or internal resistances as shown in Table 8.2. Sulfate and thiosulfate were not detected after the treatment.

The electrical charge value was determined by extrapolating the area under the current/operating time curve. This value can be used for the calculation of the amount of sulfur produced or sulfide removed as equation 8.7:

$$\text{Elemental sulfur production, } g = \frac{\text{Electrical charge} \times 32}{2F} \qquad (8.7)$$

where F is the Faraday constant (96485 coulomb mol^{-1}).

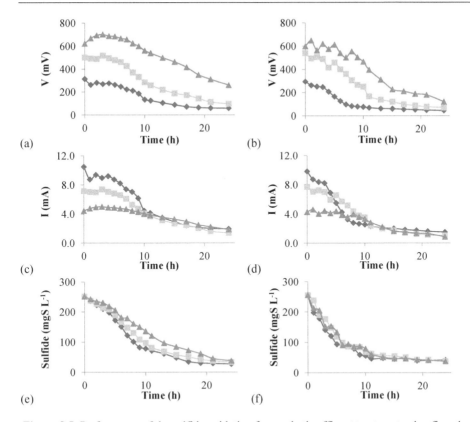

Figure 8.5. Performance of the sulfide oxidation for synthetic effluent treatment using five plus electrodes as a function of operation time (Left: pH 10 and Right: pH 7): (a, b) voltage, (c, d) current and (e, f) sulfide. (♦) 30, (■) 70 and (▲) 140 Ω.

8.3.3 Performance of five plus galvanic cells at different pH values of the synthetic sulfide solution

The performance of the five plus galvanic cells operated with synthetic sulfide solution of pH 7 and 10 were compared (Figure 8.5 and Table 8.2). Similar changing patterns of voltage, current and sulfide concentration over operating times were observed. However, electrical charges at pH 10 were higher than those at pH 7 in every case. This occurrence supported the theoretical aspect that sulfide oxidation in alkaline solutions should be better than in acid solutions (Table 8.2). The measured anode solution pH were 8, 5 and eventually 3 at the beginning, after 10 and 24 h operating time, respectively. Sulfate and thiosulfate were not detected.

8.3.4 Performance of five plus galvanic cells in treatment of real effluent

When the real effluent was applied to the five plus cell with 30 Ω external resistance, at the initial pH of 8.5 and adjusted to pH 10, results as shown in Figure 8.6 and Table 8.2 were obtained. Similar variation patterns of voltage, current and sulfide concentration over operating times were also observed. Besides, the electrical charge gained with the system operated at pH 10 was just slightly higher than those obtained with the initial effluent. The highest sulfide removal efficiencies were found with these real effluents when compared with the synthetic sulfide solution. Sulfate in the final solution was about 10 mg L^{-1} which is higher than the initial value. Characteristics of the real effluent before and after electrochemical treatment are presented in Table 8.3.

Table 8.2. Electrical charge and sulfide removal efficiency achieved in this study

Type of electrode	External resistance (Ω)	Electrical charge (I×t) (coulomb)	S^0 production (mg)	exhausted time (h)	Sulfide concentration (mg L^{-1}) Initial	Remained	Sulfide removal efficiency (%)
Synthetic sulfide wastewater with buffer pH 10							
Single rod	30	14.6	2.43	15	253	100	60.5
	70	11.9	1.98	15	252	102	59.5
Five rods	30	71.0	11.8	19	253	47.6	81.2
	70	71.8	11.9	19	256	52.5	79.5
	140	60.2	9.98	19	253	41.4	83.6
Five plus	30	431	71.4	20	253	26.8	89.4
	70	353	58.5	20	251	33.4	86.7
	140	316	52.4	20	251	37.6	85.0
Synthetic sulfide wastewater with buffer pH 7							
Five plus	30	300	49.8	13	254	41.9	83.5
	70	291	48.3	13	255	38.6	84.9
	140	231	38.3	13	256	37.4	85.4
Real sulfide wastewater with buffer pH 10							
Five plus	30	464	76.9	15	254	21.6	91.5
Real sulfide wastewater at the original pH (8.5)							
Five plus	30	407	67.5	15	253	22.6	91.1

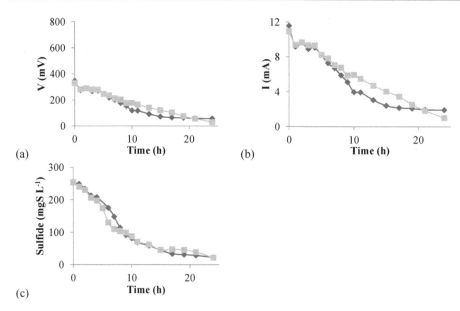

(a) (b) (c)

Figure 8.6. Performance of the sulfide oxidation for real effluent treatment using five plus electrodes and the external resistance of 30 Ω as a function of operation time: (a) voltage, (b) current and (c) sulfide. (♦) real effluent pH 10 and (■) real effluent pH 8.5.

Table 8.3. Characteristics of real effluent before and after electrochemical treatment

Parameter	Initial effluent	Treated effluent at pH 10	Treated effluent at uncorrected pH
pH	8.4 ± 0.2	8.0 ± 0.4	6.8 ± 0.3
Sulfate, mg L^{-1}	115 ± 9	125 ± 3	123 ± 6
Sulfide, mg S L^{-1}	254 ± 12	21.6 ± 2	22.6 ± 3
Calcium, mg L^{-1}	240 ± 34	112 ± 10	240 ± 29
Acetate, mg L^{-1}	229 ± 33	220 ± 15	209 ± 23

8.4. Discussion
8.4.1 The effect of internal and external resistances on an electrochemical sulfide oxidation/vanadium(V) reduction cell efficiency

This study showed that up to 91% of the sulfide can be removal by using a spontaneous electrochemical sulfide oxidation/vanadium(V) reduction process using an external resistance at 30 Ω and sulfide concentration of 250 mg L^{-1}. An enhanced electrode surface area by increasing the numbers of the electrodes or by adding graphite powder can effectively reduce the internal resistance as well as overpotential. The rate of the reaction can be controlled by changing the external resistance, as it determines the anode potential for a given current (Dutta et al., 2008). If the external resistance is too low, the sulfide oxidation rate will be high and a lack of electron transfer can occur because most of the power output of the voltage source is dissipated as heat inside the source itself (Fitzpatrick, 2007). Meanwhile the sulfide oxidation rate will be slow when a high external resistance is applied. However, in case of using 1 and 5 graphite rods, the internal resistance was very high when compared with the range of the applied external resistance. Hence, no significant differences in current or sulfide removal efficiencies were observed at each external resistance value investigated. Therefore, a high surface area electrode is required to minimize the internal resistance and overpotential as much as possible.

8.4.2. The effect of the pH on sulfide removal in an electrochemical sulfide oxidation/vanadium(V) reduction cell

According to Equation 8.2 and Figure 8.1, once an electron is transferred, cations in the anode solution will equivalently pass through the cation exchange membrane to the cathode part. Normally protons (H^+) will be the ones which move to the cathode part. However, in the case of sulfide in the pH 10 buffer solution, the H^+ formation is much lower when compared to the H^+ concentration contained in the cathode part. Thus sodium ions, which are the major cations existing in the anode solution, possibly move through the cathode part instead of H^+. This results in a decreasing of anode solution pH over operating time. Although sulfide oxidation can be occur either in acid or base conditions (Equations 8.1 and 8.2), the pH can affect the dissolution of the sulfide in the reactor. Whenever the pH of this solution decreases, hydrogen sulfide (H_2S) formation possibly occurs (Sawyer et al., 2003). Since the systems were not completely airtight, contact between the solution and ambient air allowed some H_2S loss from the anode solution. The loss of H_2S results in a reduction of the pH, thus introducing a higher rate of H_2S formation. Hence, sulfide removal efficiencies obtained in all tests were much higher than those calculated from the electrical charge values. The sulfide removal rate at pH 7 was higher than those of pH 10 (data not shown). This confirmed that conversion of sulfide ion to H_2S was occurred severely in the lower pH solution.

8.4.3. Treatment of real effluent using an electrochemical sulfide oxidation/vanadium(V) reduction cell

The sulfide removal efficiency of the real effluent was slightly higher than that achieved with synthetic sulfide wastewater. The presence of acetate does not affect sulfide oxidation (Dutta et al., 2008). However, not only sulfate reducing bacteria contained in the effluent, but other anaerobic bacteria and sulfide oxidizing bacteria are present. Therefore, not only H_2S formation can take place, biooxidation possibly also happened due to the contact of the real effluent with the oxygen in the air during operation.

As vanadium is environmentally friendly and less toxic than other electrolytes such as zinc bromine, polysulfide bromide and cerium zinc (Blanc & Rufer, 2010), it is worldwide used in metal industries, sulfuric acid production, vanadium battery manufacture, etc. Therefore, the use of vanadium in sulfide oxidation is clearly suitable for practical application. Moreover, recovery of vanadium(V) could be achieved by the oxidation of the vanadium(IV) containing cathode effluent, thus offering an almost unlimited capacity fuel cell.

Further research with completely air tight and continuous systems is required. Since closed systems not only minimize the H_2S loss, the biooxidation is also limited. More electric power production is thus expected. When the system is operated with a completely mixed reactor, variation of the sulfide concentration in the anode solution will be less. Electrical current will thus be generated constantly as sulfide oxidation will become a zero order reaction.

The decrease in electrochemical activity over time due to the deposition of elemental sulfur on the graphite electrode was found to be a major limitation of the method (Dutta et al., 2008). In order to remove solid sulfur from the electrode, the reduction of elemental sulfur to a concentrated polysulfide solutions was suggested by Dutta et al. (2009). Then, concentrated polysulfide solutions can be converted back to elemental sulfur as a solid product, either by adjusting the pH to near neutral or lightly aerating the solution. However, further studies on the regeneration of the electrode are necessary.

8.5. Conclusions

This investigation demonstrated that a spontaneous electrochemical sulfide oxidation/vanadium(V) reduction with graphite electrodes can be used for the treatment of dissolved sulfide present in the effluent of the biological sulfate reducing reactor. A high surface area electrode is required in order to minimize internal resistance and overpotential as much as possible. A sulfide removal efficiency up to 91% was achieved from real effluent treatment with an external resistance of 30 Ω at a sulfide concentration of 250 mg L^{-1}.

8.6 References

Ángela, A., Ane, U., & Inmaculada, O. (2009). Contributions of electrochemical oxidation to waste-water treatment: fundametals and review of applications. *J. Chem. Technol. Biotechnol., 84*(12), 1747-1755.

Blanc, C., & Rufer, A. (2010). Understanding the Vanadium Redox Flow Batteries. In J. Nathwani & A. Ng (Eds.), *Path to Sustainable Energy*: InTech.

Dutta, P. K., Rabaey, K., Yuan, Z., & Keller, J. (2008). Spontaneous electrochemical removal of aqueous sulfide. *Water Res., 42*, 4965-4975.

Dutta, P. K., Rozendal, R. A., Yuan, Z., Korneel, R., & Keller, J. (2009). Electrochemical regeneration of sulfur loaded eletrodes. *Electrochem. Commun., 11*, 1437-1440.

Eaton, A. D., APHA, AWWA, & WEF. (2005). *Standard methods for the examination of water and wastewater* (21st ed.). Washington D.C.

Firer, D., Friedler, E., & Lahav, O. (2008). Control of sulfide in sewer system by dosage of iron salts: comparision between theoretical and experiment results, and practical. *Sci. Total Environ., 392*(1), 145-156.

Fitzpatrick, R. (2007). Power and Internal Resistance *Electromagnetism and Optics: An introductory course*: The University of Texas at Austin.

Gabriel, D., & Deshusses, M. A. (2003). Performance of a full-scale biotrickling filter treating H$_2$S at a gas contact time of 1.6 to 2.2 s. *Environ. Prog., 22*(2), 111-118.

González-Sánchez, A., & Revah, S. (2007). The effect of chemical oxidation on the biological sulfide oxidation by an alkaliphilic sulfoxidizing bacterial consortium. *Enzyme Microb. Technol., 40*, 292-298.

González-Sánchez, A., & Revah, S. (2009). Biological sulfide removal under alkaline and aerobic conditions in a packed recycling reactor. *Water Sci. Technol., 59*(7), 1415-1421.

Henshaw, P. F., & Zhu, W. (2001). Biological conversion of hydrogen sulphide to elemental sulphur in a fixed-film continuous flow photo-reactor. *Water Res., 35*(15), 3605-3610.

Janssen, A. J. H., Sleyster, R., van der Kaa, C., Jochemsen, A., Bontsema, J., & Lettinga, G. (1995). Biological sulphide oxidation in a fed-batch reactor. *Biotechnol. Bioeng., 47*(3), 327-333.

Kijjanapanich, P., Annachhatre, A. P., Esposito, G., van Hullebusch, E. D., & Lens, P. N. L. (2013a). Biological sulfate removal from gypsum contaminated construction and demolition debris. *J. Environ. Manage., 131*, 82-91.

Kijjanapanich, P., Annachhatre, A. P., & Lens, P. N. L. (2013b). Biological sulfate reduction for treatment of gypsum contaminated soils, sediments and solid wastes. *Crit. Rev. Environ. Sci. Technol., In Press*.

Kijjanapanich, P., Do, A. T., Annachhatre, A. P., Esposito, G., Yeh, D. H., & Lens, P. N. L. (2013c). Biological sulfate removal from construction and demolition debris leachate: Effect of bioreactor configuration. *J. Hazard. Mater., In Press*.

Klymenko, O. V., & Kulys, J. (2008). Numerical simulation of electrochemical processes at a tubular electrode. Application to spectroscopy. *Nonlinear Anal. Model. Control, 13*(2), 191-199.

Kraakman, N. J. R. (2003). Robustness of a full-scale biological system treating industrial CS$_2$ emissions. *Environ. Prog., 22*(2), 79-85.

Krishnakumar, B., Majumdar, S., Manilal, V. B., & Haridas, A. (2005). Treatment of sulphide containing wastewater with sulphur recovery in a novel reverse fluidzed loop reactor (RFLR). *Water Res., 39*(4), 639-647.

Lens, P. N. L., Vallero, M., Esposito, G., & Zandvoort, M. (2002). Perspectives of sulfate reducing bioreactor in environmental biotechnology. *Rev. Environ. Sci. Biotechnol., 1*, 311-325.

Nielsen, A. H., Hvitved-Jacobsen, T., & Vollertsen, J. (2008). Effects of pH and iron concentration on sulfide precipitation in wastewater collection systems. *Water Environ. Res., 80*(4), 380-384.

Pikaar, I., Li, E., Rozendal, R. A., Yuan, Z., Keller, J., & Rabaey, K. (2012). Long-term field test of an electrochemical method for sulfide removal from sewage. *Water Res., 46*, 3085-3093.

Pikaar, I., Rozendal, R. A., Yuan, Z., Keller, J., & Rabaey, K. (2011). Electrochemical sulfide removal from synthetic and real domestic wastewater at high current densities. *Water Res., 45*, 2281-2289.

Rajeshwar, K., Ibanez, J. G., & Swain, G. M. (1994). Electrochemistry and the environment. *J. Appl. Electrochem., 24*(11), 1077-1091.

Ruwisch, R. C. (1985). A qiuck method for the determination of dissolved and precipitated sulfides in cultures of sulfate-reducing bacteria. *J. Microbiol. Methods, 4*, 33-36.

Sahinkaya, E., Hasar, H., Kaksonen, A. H., & Rittmamm, B. E. (2011). Performance of a sulfide-oxidizing, sulfur-producing membrane biofilm reactor treating sulfide-containing bioreactor effluent. *Environ. Sci. Technol., 45*(9), 4080-4087.

Sawyer, C. N., McCarty, P. L., & Parkin, G. F. (2003). *Chemistry for environmental engineering and science* (5th ed.): Mc Graw-Hill International

Smolders, A., & Roelofs, J. G. M. (1993). Sulphate-mediated iron limitation and eutrophication in aquatic ecosystems. *Aquat. Bot., 46*, 247-253.

Syed, M., Soreanu, G., Falletta, P., & Beland, M. (2006). Removal of hydrogen sulfide from gas streams using biological processes-a review. *Can. Biosyst. Eng., 48*, 2.1-2.14.

van den Ende, F. P., Laverman, A. M., & van Gemergen, H. (1996). Coexistence of aerobic chemotrophic and anaerobic phototrophic sulfur bacteria under oxygen limitation. *FEMS Microbiol. Ecol., 19*(3), 141-151.

Vincke, E., Boon, N., & Verstraete, W. (2001). Analysis of the microbial communities on corroded concrete sewer pipes - a case study. *Appl. Microbiol. Biotechnol., 57*, 776-785.

Zhang, L., Keller, J., & Yuan, Z. (2009). Inhibition of sulfate-reducing and methanogenic activites of anaerobic sewer biofilms by ferric iron dosing. *Water Res., 43*(17), 4123-4132.

CHAPTER 9

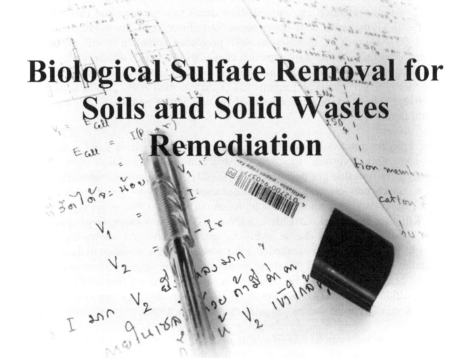

Biological Sulfate Removal for Soils and Solid Wastes Remediation

Chapter 9

Gypsiferous soils and gypsum contaminated solid wastes contain elevated concentrations of sulfate. They cause several agricultural and environmental problems such as low water retention capacity and odor problems. Reduction of the sulfate content of these gypsiferous soils and solid wastes is an option to overcome these problems. Biological sulfate reduction processes can be used to develop novel alternative techniques for treating gypsiferous soils or gypsum contaminated solid wastes. This opens perspectives for the agricultural utilization of gypsiferous soils and the decrease of the amount of gypsum contaminated solid waste. Sulfate removal is not only capable of solving these problems, but the sulfide produced in this process can also be recovered as elemental sulfur or sulfuric acid.

9.1. Introduction

Soils containing a high gypsum content (>10%) (Verheye & Boyadgiev, 1997), namely gypsiferous soils (FAO, 1990), have several problems during agricultural development such as low water retention capacity, shallow depth to a hardpan and vertical crusting (FAO, 1990). Gypsiferous soils cover about 94 million ha of the world's arable lands (FAO, 1993). Moreover, gypsiferous soil problems occur in some mining areas (Kijjanapanich et al., 2013b; Kijjanapanich et al., 2013d), which often couples with acid mine drainage (AMD) formation, thus causing a significant environmental threat, such as mass mortalities of plants and aquatic life (Kijjanapanich et al., 2013b; Kijjanapanich et al., 2012). This soil type cannot be used for agriculture and these soils have a poor fauna and flora.

Another problem related to gypsum contamination occurs with solid wastes. Solid wastes containing gypsum, such as construction and demolition debris (CDD), phosphogypsum and flue gas desulfurization (FGD) gypsum are an important source of pollution, which can create a lot of environmental problems (Azabou et al., 2007; Delaware Solid Waste Authority, 2008; Kaufman et al., 1996; Kijjanapanich et al., 2013c; Kijjanapanich et al., 2013d; U.S.EPA., 2008). Nowadays, large quantities of these wastes are generated due to industrial growth. Table 9.1 shows the generated amounts of these gypsum contaminated solid wastes and the possible toxic compounds contained in it. It is suggested that these wastes have to be separated from other wastes, especially organic waste, and be placed in a specific area of the landfill to prevent biogenic sulfide formation (Montero et al., 2010). This results in a rapid rise in the costs of the disposal of gypsum wastes (Gypsum Association, 1992).

Although these wastes can be reused as soil amendment or to make building materials, a concern has been raised by regulators regarding the chemical composition of the solid waste materials and the potential risks to human health and the environment, due to elevated concentrations of sulfate, fluoride, radioactive compounds, polycyclic aromatic hydrocarbons (PAHs) and heavy metals contained in these materials (Battistoni et al., 2006; Jang & Townsend, 2001a; Jang & Townsend, 2001b).

Reduction of the sulfate content will contribute to the utilization of gypsum contaminated soils and solid wastes (Kijjanapanich et al., 2013c). A biological sulfate reduction treatment is an attractive alternative for sulfate removal for these soils and solid wastes. In the past, biological sulfate reduction has been considered as undesirable in anaerobic wastewater treatment (Hulshoff Pol et al., 1998). In contrast, nowadays interest has grown in applying biological sulfate reduction for the treatment of specific waste streams (inorganic sulfate rich wastewaters), such as AMD or wastewater containing sulfuric acid (Kijjanapanich et al., 2012; Sahinkaya et al., 2011a), which is often coupled to heavy metal removal (Jong & Parry, 2003; Kijjanapanich et al., 2012; Liamleam, 2007). However, research on biological sulfate reduction has mainly focused on the treatment of sulfate containing groundwater or wastewaters, while research on bioremediation of gypsiferous soils and solid wastes especially using sulfate reducing bacteria (SRB) is rare. This review overviews bioremediation methods of sulfate rich soils and solid wastes.

Table 9.1. The amount of the gypsum contaminated solid wastes generated and their gypsum content and possibly toxic contaminants

Type of solid waste	The amount generation	Gypsum content (%)	Toxic compound
Construction and demolition debris (CDD)	4.9 kg m^{-2} of the structure(Turley, 1998)	1.5-37 (Jang & Townsend, 2001a; Kijjanapanich et al., 2013c)	Hevay metal such as aluminium, arsenic, cadmium, chromium, copper, zinc, lead and barium (Jang & Townsend, 2001a; Kijjanapanich et al., 2013c) and organic compounds, such as toluene, trichlorofluoromethane and several PAHs (Jang & Townsend, 2001b)
Phosphogypsum	100-280 million tons per year worldwide (Tayibi et al., 2009)	Depends on the phosphate rock source material, can be up to 90% (Rutherford et al., 1994)	Residual acid, fluoride, toxic metals such as lead, selenium, strontium and Cerium (Mulopo & Ikhu-Omoregbe, 2012), and radioactive compounds such as uranium, radium and radon (Azabou et al., 2005; Rutherford et al., 1995)
Flue gas desulfurization (FGD) gypsum	-	Depends on the coal source material	Fluoride, toxic metals and radioactive compounds

9.2. Biological versus Chemical Treatment for Sulfate Removal

Sulfate removal processes can be either by biological or chemical processes (Azabou et al., 2007; Benatti et al., 2009; Dar et al., 2007; Hlabela et al., 2007). A variety of physico-chemical treatment processes are employed for sulfate removal such as ion exchange, adsorption and membrane filtration. These technologies are, however, relatively expensive due to their higher operation, maintenance costs and energy consumption (Ozacar et al., 2008). Chemical precipitation is a well-established technology with ready availability of equipment and chemicals (U.S.EPA, 2000). Barium (Ba) and Lead (Pb) compounds, such as $BaCl_2$ and $Pb(NO_3)_2$, are well-known efficient chemicals for sulfate removal (Benatti et al., 2009; Maree et al., 2004) with a sulfate removal efficiency up to 90% (Bosman et al., 1990; Hlabela et al., 2007; Kijjanapanich et al., 2013a).

Although Ba^{2+} and Pb^{2+} compounds show good performance in sulfate precipitation, residual Ba^{2+} and Pb^{2+} which remains in the treated leachate or material after the precipitation process are toxic (Benatti et al., 2009). They can result in an adverse impact on the environment if these are discharged or used without any post-treatment. Calcium compounds can be cheap alternative chemicals for sulfate removal, as these are less toxic than Ba and Pb. Calcium compounds, such as calcium chloride and calcium oxide, showed a good sulfate removal efficiency in many studies (Benatti et al., 2009; Bosman et al., 1990; Hlabela et al., 2007; Maree et al., 2004). However, if calcium was used as precipitant, a residual sulfate concentration up to 1450 mg L^{-1} of sulfate will remain due to the high solubility of calcium sulfate (gypsum). In addition, systems for precipitate separation and appropriate reuse or disposal of the solid phase are necessary when using chemical sulfate removal processes (Silva et al., 2002). Box 9.1 summarizes the advantages and disadvantages of the chemical sulfate removal.

Avantages of chemical sulfate removal
- High sulfate removal efficiency
- Require short treatment times
- Require small reactor volume
- No need for a sophisticated operation
- Low maintenance costs (requiring only replenishment of the chemicals used)

Disavantages of chemical sulfate removal
- Expensive chemicals
- Remaining of toxic chemical in the treated water
- Require liquid-solid separation system

Box 9.1. Summary of the advantages and disadvantages of the chemical sulfate removal.

A biological sulfate reduction system makes use of the bacterial sulfate reduction process as it occurs in nature for the removal of sulfate, often coupled to heavy metal removal (Jong & Parry, 2003; Kijjanapanich et al., 2012; Liamleam, 2007). The biological sulfate reduction approach involves the use of anaerobic SRB, which reduces sulfate to sulfide by oxidizing an organic carbon source (Equation 9.1):

$$2CH_2O + SO_4^{2-} + 2H^+ \rightarrow H_2S + 2CO_2 + H_2O \tag{9.1}$$

where CH_2O represents a simple organic compound. The addition of an electron donor, such as ethanol or lactate is necessary in case of biological sulfate reduction (Liamleam & Annachhatre, 2007). However, low or no cost organic substrates, such as wood chips, compost, and sewage sludge, can also be used (Gibert et al., 2004; Waybrant et al., 1998). These organic substrates are much cheaper and less toxic when compared to bulk chemicals. Box 9.2 summarizes the advantages and disadvantages of the biological sulfate reduction process.

9.3. Biological Sulfate Removal for Soils Treatment

Research on bioremediation of gypsiferous soils, especially using SRB, is rare. Soils containing significant quantities of gypsum, which may interfere with plant growth, are defined as gypsiferous soils (FAO, 1990). Most of the gypsiferous soils have a relatively low organic matter content (Ghabour et al., 2008). Therefore, sufficient electron donor for SRB needs to be supplied when the soils are treated by biological sulfate reduction (Alfaya et al., 2009; Kijjanapanich et al., 2013b).

Avantages of biological sulfate reduction
- Both sulfate and metals can be reduced to very low levels
- The amount of waste produced is minimal
- Capital costs are relatively low
- Operating costs can be drasticaly reduced by using no or low cost electron donor and carbon sources
- Less toxic compounds produced

Disavantages of biological sulfate reduction
- Slow process kinetics
- Requirement and cost of an external electron donor
- Need for a post-treatment of the sulfide containing effluent

Box 9.2. Summary of the advantages and disadvantages of the biological sulfate reduction process.

A bioremediation technology to remove the gypsum content of gypsiferous soils by SRB was developed (Figure 9.1) (Alfaya et al., 2009; Kijjanapanich et al., 2013b). Alfaya et al. (2009) found that the calcareous gypsiferous soils from Spain contained an endogenous SRB population that uses the sulfate from gypsum in the soil as electron acceptor. However, the sulfate reduction rate doubled when anaerobic granular sludge was added to bioaugment the soil with SRB. In the presence of anaerobic granular sludge, a maximum sulfate reduction rate of 567 mg L^{-1} d^{-1} was achieved with propionate as the electron donor (Alfaya et al., 2009).

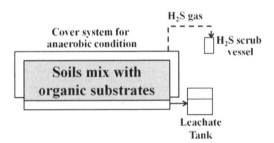

Figure 9.1. Treatment concept for gypsum contaminated soils

The cheap organic substrates, such as rice husk (RH), coconut husk chip (CHC) and pig farm wastewater treatment sludge (PWTS), which are suggested for use as electron donor in permeable reactive barrier (PRB) systems (Kijjanapanich et al., 2012), can be also used as electron donor for SRB in biological sulfate reduction for gypsiferous soils treatment (Kijjanapanich et al., 2013b). The combination of PWTS, RH and CHC was used for the treatment of the gypsiferous soils in the study of Kijjanapanich et al. (2013b). The gypsum mine soil (overburden) from Thailand was mixed with an organic mixture in different amounts. The highest sulfate removal efficiency (59%) was achieved in the soil mixture which contained 40% organic material, corresponding to a reduction of the soil gypsum content from 25% to 7.5%. The organic matter is not only used as electron donor for the SRB, but can also be as nutrient source for the plant (Kijjanapanich et al., 2013b).

9.4. Biological Sulfate Removal for Solid Wastes Treatment

Gypsum contaminated solid wastes, such as CDD, phosphogypsum and FGD gypsum can be treated by biological sulfate reduction (Castillo et al., 2012; Kijjanapanich et al., 2013c; Wolicka & Borkowski, 2009; Zhao et al., 2010). Sulfate contained in these solid wastes was shown to be a good source of sulfate for SRB in many studies (Rzeczycka et al., 2004; Wolicka & Kowalski, 2006). Similar as in gypsiferous soils, most of these solid wastes have a low organic carbon content and additional of an electron donor is necessary (Rzeczycka et al., 2001).

Treatment of these solid wastes can be done in two ways (Figure 9.2), including an indirect (Figure 9.3a) or a direct (Figure 9.3b) treatment. In the indirect treatment concept, the gypsum contained in the CDD is leached out by water in a leaching step. The sulfate containing leachate is further treated in a biological sulfate reduction system. The treated water from the bioreactor can then be reused in the leaching column (Figure 9.3a). The leaching step was found to be the most time consuming step for this kind of treatment (Kijjanapanich et al., 2013c). Kijjanapanich et al. (2013c) found that the treated CDD 0.3-0.7 g sulfate kg^{-1} sand, which is far below the Dutch government limit for the maximum amount of sulfate present in building sand and could be reused in construction activities.

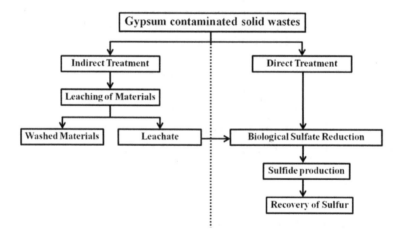

Figure 9.2. Treatment concept for gypsum contaminated solid waste.

In the direct treatment concept, the solid wastes are directly mixed with the electron donor in the bioreactor (Hiligsmann et al., 1996; Kaufman et al., 1996; Kijjanapanich et al., 2013c) (Figure 9.3b). This depends on the content of gypsum in the solid wastes (Table 9.1). The sulfide produced from this biological process can be recovered as elemental sulfur (S^0) (Dutta et al., 2008; Pikaar et al., 2011; Sahinkaya et al., 2011b) or sulfuric acid (H$_2$SO$_4$) (Laursen & Karavanov, 2006).

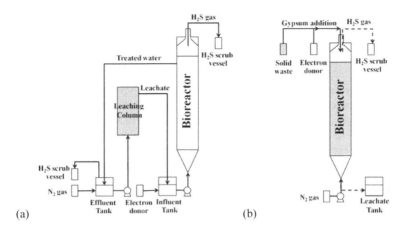

Figure 9.3. Treatment for gypsum contaminated solid waste: (a) indirect treatment and (b) direct treatment. (■) indicates the location of the solid wastes in the system.

9.5. Future Perspectives

The remediation of gypsiferous soils by a biological sulfate reducing process can be applied for either *ex situ* or *in situ* gypsiferous soils treatment. However, further studies of treating gypsiferous soils from different sources using this biological sulfate reduction system are recommended to compare and investigate the effect of the soil composition on the sulfate reduction process. Moreover, further studies are nevertheless necessary to explore the agricultural potential of the treated soils.

In practice, direct recovery of sulfur from the gas phase may be complicated and difficult, especially in case of the *in situ* treatment which normally covers enormous areas. Therefore, recovery of sulfur from sulfide contained in the leachate of the system can be an alternative option. A high sulfide accumulation can also be achieved in the reactor treating solid wastes containing gypsum, especially in full scale applications where a high sulfate loading rate is applied. The removal of sulfide from the system as well as the effluent of the biological sulfate reduction process is required, as sulfide can give an adverse effect to SRB in the system (Al-Zuhair et al., 2008) and cause several environmental impacts (Lens & Kuenen, 2001; Vincke et al., 2001) or be re-oxidized to sulfate if directly discharged into the environment.

The development of the biological sulfate reduction process combined with a sulfur recovery system has to be achieved. Either chemical or biological processes can be applied for sulfide oxidation to elemental sulfur (González-Sánchez & Revah, 2007). Nowadays, a biological sulfide oxidation using oxygen as electron acceptor and sulfide oxidizing bacteria as a catalyst is a very popular system (González-Sánchez & Revah, 2009; Henshaw & Zhu, 2001; Krishnakumar et al., 2005; Sahinkaya et al., 2011b). However, this system requires energy for oxygen supply (Syed et al., 2006; van den Ende et al., 1996), complicated operation techniques (Syed et al., 2006) and the pH conditions in these biological systems are usually mildly or extremely acidic (Gabriel & Deshusses, 2003; Kraakman, 2003). Oversupply of oxygen also yields a lower sulfate removal efficiency since most sulfide is changed to sulfate instead of sulfur (Janssen et al., 1995). Electrochemical treatment of such wastewaters can be an appropriate way which offers several advantages, including good energetic efficiency, environmental compatibility, versatility, selectivity and cost effectiveness (Ángela et al., 2009; Dutta et al., 2009), especially a spontaneous reaction or a galvanic cell, which not only removes sulfide, but sulfur can be recovered as elemental sulfur and electricity can be generated.

At the higher concentrations of solid wastes containing gypsum, SRB growth could be inhibited (Azabou et al., 2005; Rzeczycka et al., 2004), due to an accumulation of toxic levels of impurities, especially fluorine and heavy metals (Kijjanapanich et al., 2013c; Rzeczycka et al., 2004). Heavy metals, such as aluminium, arsenic, cadmium, chromium and copper, can inhibit the growth rate of SRB (Azabou et al., 2005; Rzeczycka et al., 2004; Townsend et al., 2004), depending on their speciation and concentration. Further studies are needed to reduce the toxicity of metals, radioactive and PAH compounds present in these solid wastes to SRB.

In order to control the formation of desirable end products in sulfate reduction systems, process control which has been used for several biological production processes can be an alternative option (Dunn et al., 2005; Villa-Gomez et al., 2013). With better process control, excess sulfide or COD formation can be avoided, thus decreasing of the operational cost and eliminating the need for a post-treatment step.

A calcium recovery step might be required in the process to prevent accumulation of calcium carbonate precipitates in the piping or the granular sludge, e.g. by microbial carbonate precipitation (MCP) using ureolytic bacteria (Al-Thawadi & Cord-Ruwisch, 2012; Al-Thawadi, 2011; Hammes et al., 2003; Whiffin et al., 2007).

9.6 References

Al-Thawadi, S., & Cord-Ruwisch, R. (2012). Calcium carbonate crystals formation by ureolytic bacteria isolated from Australian soil and sludge. *J. Adv. Sci. Eng. Res., 2*, 12-26.

Al-Thawadi, S. M. (2011). Ureolytic bacteria and calcium carbonate formation as a mechanism of strength enhancement of sand. *J. Adv. Sci. Eng. Res., 1*, 98-114.

Al-Zuhair, S., El-Naas, M. H., & Al-Hassani, H. (2008). Sulfate inhibition effect on sulfate reducing bacteria. *J. Biochem. Technol., 1*(2), 39-44.

Alfaya, F., Cuenca-Sánchez, M., Garcia-Orenes, F., & Lens, P. N. L. (2009). Endogenous and bioaugmented sulphate reduction in calcareous gypsiferous soils. *Environ. Technol., 30*(12), 1305-1312.

Ángela, A., Ane, U., & Inmaculada, O. (2009). Contributions of electrochemical oxidation to waste-water treatment: fundametals and review of applications. *J. Chem. Technol. Biotechnol., 84*(12), 1747-1755.

Azabou, S., Mechichi, T., & Sayadi, S. (2005). Sulfate reduction from phosphogypsum using a mixed culture of sulfate-reducing bacteria. *Int. Biodeter. Biodegr., 56*(4), 236-242.

Azabou, S., Mechichi, T., & Sayadi, S. (2007). Zinc precipitation by heavy-metal tolerant sulfate-reducing bacteria enriched on phosphogypsum as a sulfate source. *Miner. Eng., 20*, 173-178.

Battistoni, P., Carniani, E., Fatone, F., Balboni, P., & Tornabuoni, P. (2006). Phosphogypsum leachate: Treatment feasibility in a membrane plant. *Ind. Eng. Chem. Res., 45*, 6504-6511.

Benatti, C. T., Tavares, C. R. G., & Lenzi, E. (2009). Sulfate removal from waste chemicals by precipitation. *J. Environ. Manage., 90*, 504-511.

Bosman, D. J., Clayton, J. A., Maree, J. P., & Adlem, C. J. L. (1990). Removal of sulphate from mine water with barium sulphide. *Mine Water Environ., 9*, 149-163.

Castillo, J., Pérez-López, R., Sarmiento, A. M., & Nieto, J. M. (2012). Evaluation of organic substrates to enhance the sulfate-reducing activity in phosphogypsum. *Sci. Total Environ., 439*, 106-113.

Dar, S. A., Stams, A. J., Kuenen, J. G., & Muyzer, G. (2007). Co-existence of physiologically similar sulfate-reducing bacteria in a full-scale sulfidogenic bioreactor fed with a single organic electron donor. *Appl. Microbiol. Biotachnol., 75*(6), 1463-1472.

Delaware Solid Waste Authority. (2008). Construction and demolition (C&D) debris. Retrieved January 16, 2011, from http://www.dswa.com/programs_construction.asp

Dunn, I. J., Heinzle, E., Ingham, J., & Přenosil, J. E. (2005). *Biological Reaction Engineering*: Wiley-VCH Verlag GmbH & Co. KGaA.

Dutta, P. K., Rabaey, K., Yuan, Z., & Keller, J. (2008). Spontaneous electrochemical removal of aqueous sulfide. *Water Res., 42*, 4965-4975.

Dutta, P. K., Rozendal, R. A., Yuan, Z., Korneel, R., & Keller, J. (2009). Electrochemical regeneration of sulfur loaded eletrodes. *Electrochem. Commun., 11*, 1437-1440.

FAO. (1990). *FAO Soils Bulletin 62: Management of gypsiferous soils*. Rome.

FAO. (1993). *World Soil Resources Report 66: An explanation of the FAO World Soil Resources map at a scale of 1:25 000 000*. Rome.

Gabriel, D., & Deshusses, M. A. (2003). Performance of a full-scale biotrickling filter treating H_2S at a gas contact time of 1.6 to 2.2 s. *Environ. Prog., 22*(2), 111-118.

Ghabour, T. K., Aziz, A. M., & Rahim, I. S. (2008). Anthropogenic impact of fertilization on gypsiferous soils. *Am. Eurasian J. Agric. Environ. Sci., 4*(4), 405-409.

Gibert, O., de Pablo, J., Cortina, J. L., & Ayora, C. (2004). Chemical characterization of natural organic substrates for biological mitigation of acid mine drainage. *Water Res., 38*, 4186-4196.

González-Sánchez, A., & Revah, S. (2007). The effect of chemical oxidation on the biological sulfide oxidation by an alkaliphilic sulfoxidizing bacterial consortium. *Enzyme Microb. Technol., 40*, 292-298.

González-Sánchez, A., & Revah, S. (2009). Biological sulfide removal under alkaline and aerobic conditions in a packed recycling reactor. *Water Sci. Technol., 59*(7), 1415-1421.

Gypsum Association. (1992). Treatment and disposal of gypsum board waste: Technical paper part II, *AWIC's Construction Dimensions* (Vol. March): AWIC.

Hammes, F., Seka, A., de Knijf, S., & Verstraete, W. (2003). A novel approach to calcium removal from calcium-rich industrial wastewater. *Water Res., 37*(3), 699-704.

Henshaw, P. F., & Zhu, W. (2001). Biological conversion of hydrogen sulphide to elemental sulphur in a fixed-film continuous flow photo-reactor. *Water Res., 35*(15), 3605-3610.

Hiligsmann, S., Deswaef, S., Taillieu, X., Crine, M., Milande, N., & Thonart, P. (1996). Production of sulfur from gypsum as an industrial by-product. *Appl. Biochem. Biotechnol., 57-58*, 959-969.

Hlabela, P., Maree, J., & Bruinsma, D. (2007). Barium carbonate process for sulphate and metal removal from mine water. *Mine Water Environ., 26*, 14-22.

Hulshoff Pol, L. W., Lens, P. N. L., Stams, A. J. M., & Lettinga, G. (1998). Anaerobic treatment of sulphate-rich wastewaters. *Biodegrad., 9*, 213-224.

Jang, Y. C., & Townsend, T. (2001a). Sulfate leaching from recovered construction and demolition debris fines. *Adv. Environ. Res., 5*, 203-217.

Jang, Y. C., & Townsend, T. G. (2001b). Occurrence of organic pollutants in recovered soil fined from construction and demolition waste. *Waste Manage., 21*, 703-715.

Janssen, A. J. H., Sleyster, R., van der Kaa, C., Jochemsen, A., Bontsema, J., & Lettinga, G. (1995). Biological sulphide oxidation in a fed-batch reactor. *Biotechnol. Bioeng., 47*(3), 327-333.

Jong, T., & Parry, D. L. (2003). Removal of sulfate and heavy metals by sulfate-reducing bacteria in short term bench scale upflow anaerobic packed bed reactor runs. *Water Res., 37*, 3379-3389.

Kaufman, E. N., Little, M. H., & Selvaraj, P. T. (1996). Recycling of FGD gypsum to calcium carbonate and elemental sulfur using mixed sulfate-reducing bacteria with sewage digest as a carbon source. *J. Chem. Technol. Biotechnol., 66*, 365-374.

Kijjanapanich, P., Annachhatre, A. P., Esposito, G., & Lens, P. N. L. (2013a). Chemical sulfate removal for treatment of construction and demolition debris leachate. *Environ. Technol., Submitted.*

Kijjanapanich, P., Annachhatre, A. P., Esposito, G., & Lens, P. N. L. (2013b). Use of organic substrates as electron donors for biological sulfate reduction in gypsiferous mine soils from Nakhon Si Thammarat (Thailand). *Chemosphere, In Press.*

Kijjanapanich, P., Annachhatre, A. P., Esposito, G., van Hullebusch, E. D., & Lens, P. N. L. (2013c). Biological sulfate removal from gypsum contaminated construction and demolition debris. *J. Environ. Manage., 131*, 82-91.

Kijjanapanich, P., Annachhatre, A. P., & Lens, P. N. L. (2013d). Biological sulfate reduction for treatment of gypsum contaminated soils, sediments and solid wastes. *Crit. Rev. Environ. Sci. Technol., In Press.*

Kijjanapanich, P., Pakdeerattanamint, K., Lens, P. N. L., & Annachhatre, A. P. (2012). Organic substrates as electron donors in permeable reactive barriers for removal of heavy metals from acid mine drainage. *Environ. Technol., 33*(23), 2635-2644.

Kraakman, N. J. R. (2003). Robustness of a full-scale biological system treating industrial CS$_2$ emissions. *Environ. Prog., 22*(2), 79-85.

Krishnakumar, B., Majumdar, S., Manilal, V. B., & Haridas, A. (2005). Treatment of sulphide containing wastewater with sulphur recovery in a novel reverse fluidzed loop reactor (RFLR). *Water Res., 39*(4), 639-647.

Laursen, J. K., & Karavanov, A. N. (2006). Processes for sulfur recovery, regeneration of spent acid, and reduction of NO$_x$ emissions. *Chem. Pet. Eng., 42*(5-6), 229-234.

Lens, P. N. L., & Kuenen, J. G. (2001). The biological sulfur cycle: novel opportunities for environmental biotechnology. *Water Sci. Technol., 44*(8), 57-66.

Liamleam, W. (2007). *Zinc removal from industrial discharge using thermophilic biological sulfate reduction with molasses as electron donor.* Asian Institute of Technology, Thailand.

Liamleam, W., & Annachhatre, A. P. (2007). Electron donors for biological sulfate reduction. *Biotechnol. Adv., 25*(5), 452-463.

Maree, J. P., Hlabela, P., Nengovhela, R., Geldenhuys, A. J., Mbhele, N., Nevhulaudzi, T., et al. (2004). Treatment of mine water for sulphate and metal removal using barium sulphide. *Mine Water Environ., 23,* 195-203.

Montero, A., Tojo, Y., Matsuto, T., Yamada, M., Asakura, H., & Ono, Y. (2010). Gypsum and organic matter distribution in a mixed construction and demolition waste sorting process and their possible removal from outputs. *J. Hazard. Mater., 175,* 747-753.

Mulopo, J., & Ikhu-Omoregbe, D. (2012). Phosphogypsum conversion to calcium carbonate and utilization for remediation of acid mine drainage. *J. Chem. Eng. Process. Tachnol., 3*(2).

Ozacar, M., Sengil, I. A., & Turkmenler, H. (2008). Equilibrium and kinetic data, and adsorption mechanism for adsorption of lead onto valonia tannin resin. *Chem. Eng. J., 143,* 32-42.

Pikaar, I., Rozendal, R. A., Yuan, Z., Keller, J., & Rabaey, K. (2011). Electrochemical sulfide removal from synthetic and real domestic wastewater at high current densities. *Water Res., 45,* 2281-2289.

Rutherford, P. M., Dudas, M. J., & Arocena, J. M. (1995). Radioactivity and elemental composition of phosphogypsum produced from three phophate rock sources. *Waste Manag. Res., 13,* 407-423.

Rutherford, P. M., Dudas, M. J., & Samek, R. A. (1994). Environmental impact of phosphogypsum. *Sci. Total Environ., 149,* 1-38.

Rzeczycka, M., Mycielski, R., Kowalski, W., & Galazka, M. (2001). Biotransformation of phosphogypsum in media containing defferent forms of nitrogen. *Acta Mocrobiol. Pol., 50,* 3-4.

Rzeczycka, M., Suszek, A., & Blaszczyk, M. (2004). Biotransformation of phosphogypsum by sulphate-reducing bacteria in media containing different zinc salts. *Pol. J. Environ. Stud., 13*(2), 209-217.

Sahinkaya, E., Gunes, F. M., Ucar, D., & Kaksonen, A. H. (2011a). Sulfidogenic fluidized bed treatment of real acid mine drainage water. *Bioresour. Technol., 102,* 683-689.

Sahinkaya, E., Hasar, H., Kaksonen, A. H., & Rittmamm, B. E. (2011b). Performance of a sulfide-oxidizing, sulfur-producing membrane biofilm reactor treating sulfide-containing bioreactor effluent. *Environ. Sci. Technol., 45*(9), 4080-4087.

Silva, A. J., Varesche, M. B., Foresti, E., & Zaiat, M. (2002). Sulphate removal from industrial wastewater using a packed-bed anaerobic reactor. *Process Biochem., 37,* 927-935.

Syed, M., Soreanu, G., Falletta, P., & Beland, M. (2006). Removal of hydrogen sulfide from gas streams using biological processes-a review. *Can. Biosyst. Eng., 48,* 2.1-2.14.

Tayibi, H., Choura, M., Lopez, F. A., Alguacil, F. J., & Lopez-Delgado, A. (2009). Environmental impact and management of phosphogypsum. *J. Environ. Manage., 90,* 2377-2386.

Townsend, T., Tolaymat, T., Leo, K., & Jambeck, J. (2004). Heavy metals in recovered fines from construction and demolition debris recycling facilities in Florida. *Sci. Total Environ., 332,* 1-11.

Turley, W. (1998). What's happening in gypsum recycling. *C&D Debris Recycling, 5*(1), 8-12.

U.S.EPA. (2000). *Wastewater Technology Fact Sheet Chemical Precipitation.* Washington, D.C.

U.S.EPA. (2008). Agricultural uses for flue gas desulfulfurization (FGD) Gypsum.

van den Ende, F. P., Laverman, A. M., & van Gemergen, H. (1996). Coexistence of aerobic chemotrophic and anaerobic phototrophic sulfur bacteria under oxygen limitation. *FEMS Microbiol. Ecol., 19*(3), 141-151.

Verheye, W. H., & Boyadgiev, T. G. (1997). Evaluating the land use potential of gypsiferous soils from field pedogenic characteristics. *Soil Use Manage., 13*, 97-103.

Villa-Gomez, D. K., Cassidy, J., Keesman, K., Sampaio, R., & Lens, P. N. L. (2013). Tuning strategies to control the sulfide concentration using a pS electrode in sulfate reducing bioreactor. *Water Res., Submitted*.

Vincke, E., Boon, N., & Verstraete, W. (2001). Analysis of the microbial communities on corroded concrete sewer pipes - a case study. *Appl. Microbiol. Biotechnol., 57*, 776-785.

Waybrant, K. R., Blowes, D. W., & Ptacek, C. J. (1998). Selection of Reactive Mixtures for Use in Permeable Reactive Walls for Treatment of Mine Drainage. *Environ. Sci. Technol., 32*(13), 1972-1979.

Whiffin, V. S., van Paassen, L. A., & Harkes, M. P. (2007). Microbial carbonate precipitation as a soil improvement technique. *Geomicrobiol. J., 24*, 417-423.

Wolicka, D., & Borkowski, A. (2009). Phosphogypsum biotransformation in cultures of sulphate reducing bacteria in whey. *Int. Biodeter. Biodegr., 63*, 322-327.

Wolicka, D., & Kowalski, W. (2006). Biotransformation of phosphogypsum in petroleum-refining wastewaters. *Pol. J. Environ. Stud., 15*(2), 355-360.

Zhao, Y., Chen, C., & Han, Y. (2010). Study on treating desulfurization gypsum by sulfate-reducing bacteria. *J. Environ. Technol. Eng., 3*(1), 5-10.

SENSE

Netherlands Research School for the
Socio-Economic and Natural Sciences of the Environment

C E R T I F I C A T E

The Netherlands Research School for the
Socio-Economic and Natural Sciences of the Environment
(SENSE), declares that

Pimluck Kijjanapanich

born on 1 August 1985 in Chiang Mai, Thailand

has successfully fulfilled all requirements of the
Educational Programme of SENSE.

Delft, 18 November 2013

the Chairman of the SENSE board

Prof. dr. Rik Leemans

the SENSE Director of Education

Dr. Ad van Dommelen

The SENSE Research School has been accredited by the Royal Netherlands Academy of Arts and Sciences (KNAW)

K O N I N K L I J K E   N E D E R L A N D S E
A K A D E M I E   V A N   W E T E N S C H A P P E N

The SENSE Research School declares that Ms. Pimluck Kijjanapanich has successfully fulfilled all requirements of the Educational PhD Programme of SENSE with a work load of 45 ECTS, including the following activities:

Curriculum Vitae

Pimluck Kijjanapanich was born on 1st August 1985 in Chiang Mai (Thailand). She did her primary and secondary school at Dara Academy and The Prince Royal's College in Chiang Mai (Thailand), respectively. In the years 2004-2008, she had studied Chemistry at Chiang Mai University in Chiang Mai (Thailand) for her Bachelor's degree, which was funded by The Human Development in Science Project of Thailand. She got the first class honors with silver medal in Bachelor of Science (Chemistry) in 2008. From 2008 till 2010, she got the scholarship funded by Her Majesty Queen of Thailand to do her Master of Science in Environmental Engineering and Management at Asian Institute of Technology (AIT) (Thailand). After graduation, she worked as a research associate for The Asian Regional Research Program on Environmental Technology (ARRPET) at AIT for 4 months. From November 2010, she started her PhD in the Joint Doctorate program funded by Erasmus Mundus Joint Doctorate Environmental Technologies for Contaminated Solids, Soils and Sediments (ETeCoS^3). The research was carried out at UNESCO-IHE Institute for Water Education (The Netherlands), University of Cassino and Southern Lazio (Italy) and Asian Institute of Technology (Thailand). Most of her researches are focus mainly on biological sulfate reduction for wastewater, solid waste and soils treatment.

Publications and Conferences

I. Publications

- **Kijjanapanich, P.,** Kijjanapanich, P., Annachhatre, A. P., Esposito, G., & Lens, P. N. L. (2013). Spontaneous electrochemical treatment for sulfur recovery by a sulfide oxidation/vanadium(V) reduction galvanic cell. *J. Electroanal. Chem., Submitted.*

- **Kijjanapanich, P.,** Annachhatre, A. P., Esposito, G., & Lens, P. N. L. (2013). Chemical sulfate removal for treatment of construction and demolition debris leachate. *Environ. Technol., Submitted.*

- **Kijjanapanich, P.,** Annachhatre, A. P., Esposito, G., & Lens, P. N. L. (2013). Use of organic substrates as electron donors for biological sulfate reduction in gypsiferous mine soils from Nakhon Si Thammarat (Thailand). *Chemosphere, In Press.* DOI: 10.1016/j.chemosphere.2013.11.026

- **Kijjanapanich, P.,** Do, A. T., Annachhatre, A. P., Esposito, G., Yeh, D. H., & Lens, P. N. L. (2013). Biological sulfate removal from construction and demolition debris leachate: Effect of bioreactor configuration. *J. Hazard. Mater. [G16 Special Issue], In Press.* DOI: 10.1016/j.jhazmat.2013.10.015

- **Kijjanapanich, P.,** Annachhatre, A. P., Esposito, G., van Hullebusch, E. D., & Lens, P. N. L. (2013). Biological sulfate removal from gypsum contaminated construction and demolition debris. *J. Environ. Manage., 131,* 82-91. DOI: 10.1016/j.jenvman.2013.09.025

- **Kijjanapanich, P.,** Annachhatre, A. P., & Lens, P. N. L. (2013). Biological sulfate reduction for treatment of gypsum contaminated soils, sediments and solid wastes. *Crit. Rev. Environ. Sci. Technol., In Press.* DOI: 10.1080/10643389.2012.743270

- **Kijjanapanich, P.,** Pakdeerattanamint, K., Lens, P. N. L., & Annachhatre, A. P. (2012). Organic substrates as electron donors in permeable reactive barriers for removal of heavy metals from acid mine drainage. *Environ. Technol., 33*(23), 2635-2644. DOI: 10.1080/09593330.2012.673013

II. Conferences

- **Kijjanapanich, P.,** Annachhatre, A. P., Esposito, G., & Lens, P. N. L. (2013). Biological sulfate reduction in different reactor configurations for treatment of construction and demolition debris sand leachate. *Proceedings of the 3rd IWA BENELUX Young Water Professionals Regional Conference.* Belval, Luxembourg.

- **Kijjanapanich, P.,** Annachhatre, A. P., Esposito, G., & Lens, P. N. L. (2013). Sulfate reduction for remediation of gypsiferous soils and wastes.

Paper presented at the ETeCoS³ Summer School on Contaminated Sediments: Characterization and Remediation, Delft, The Netherlands.

- **Kijjanapanich, P.**, Annachhatre, A. P., Esposito, G., van Hullebusch, E. D., & Lens, P. N. L. (2013). Biological sulfate removal from construction and demolition debris. *Proceedings of the 3rd International Conference on Research Frontiers in Chalcogen Cycle Science & Technology* (pp. 101-106). Delft, The Netherlands.

- **Kijjanapanich, P.**, Annachhatre, A. P., Esposito, G., & Lens, P. N. L. (2012). Biological sulfate reduction for remediation of gypsiferous soils using organic substrates as electron donors. *Paper presented at the SENSE Science Market: Towards a biobased economy*, Den Haag, The Netherlands.

- **Kijjanapanich, P.**, Annachhatre, A. P., Esposito, G., & Lens, P. N. L. (2012). Biological sulfate reduction for remediation of gypsiferous soils using organic substrates as electron donors. *Paper presented at the UNESCO-IHE PhD Week: Managing Water Resources in a Changing World*, Delft, The Netherlands.

- **Kijjanapanich, P.**, Annachhatre, A. P., Esposito, G., & Lens, P. N. L. (2012). Biological Sulfate Reduction for Remediation of Gypsiferous Soils and Wastes. *Paper presented at the ETeCoS³ Summer school on Contaminated Soils: from characterization to remediation*, Marne la Vallée, France.

- **Kijjanapanich, P.**, Annachhatre, A. P., & Lens, P. N. L. (2011). Treatment for Removal of Sulfate from Gypsum Contaminated Wastes. *Proceedings of EurAsia Waste Management Symposium* (pp. 572-579). Istanbul, Turkey.

- **Kijjanapanich, P.**, Annachhatre, A. P., Esposito, G., & Lens, P. N. L. (2011). Sulfate reduction for remediation of gypsiferous soils and wastes. *Paper presented at the UNESCO-IHE PhD Week: Optimizing Water Use with a Focus on Developing Countries*, Delft, The Netherlands.

- **Kijjanapanich, P.** (2011). Sulfate reduction for remediation of gypsiferous soils and wastes. *Paper presented at the ETeCoS³ Summer School: biological and thermal treatment of municipal solid waste*, Naples, Italy.

T - #0418 - 101024 - C42 - 240/170/10 - PB - 9781138015357 - Gloss Lamination